Quantum Rendering

Games, Minds, and the Future of AI

Rob Vermiller

Table of Contents

Introduction

There's a certain allure to the idea of "rendering", a concept that connects seemingly disparate realms – from video games' vivid landscapes to quantum mechanics' perplexing intricacies, the enigma of consciousness, the labyrinth of free will, and the accelerating evolution of artificial intelligence (AI). When you're fully immersed in a video game, the world seems boundless and real. Yet, only the parts of this world with which you're engaging directly truly come to life; the remainder sets in a hazy, unrealized state, like a fog waiting to be lifted. This clever trick of rendering saves computational power by bringing into focus only what's necessary at any given moment, which maintains the illusion of a complete and continuous universe. Yet, what if this principle didn't just belong in the digital realm? What if, on a grander, more profound scale, the same idea – that observation shapes reality – applies to the fabric of the universe, the workings of our minds, and even the choices that we make?

To understand this, we need to plunge into the baffling world of quantum mechanics, where particles like electrons and photons defy our classical understanding of reality. A physicist in a dimly-lit lab, surrounded by beams of light, delicate instruments, and a tangle of wires, waits eagerly until the exact moment when he finally pinpoints a particle's elusive position. Until that moment, the particle exists in a strange limbo of possibilities, like the unrendered sections of a game map laying just beyond your field of view. It's a fundamental truth about our universe that suggests that reality itself might be influenced by the act of observing it – a concept that has left scientists and philosophers scratching their heads and engaging in debates for decades.

However, this idea doesn't just hover in the lofty heights of quantum physics; it's intimately tied to the way that our minds work, as they constantly simulate and render our personal versions of reality. Sitting quietly in your favorite café, sipping coffee, your thoughts meander through imagined scenarios and potential futures. Your consciousness acts as the ultimate observer, continuously sifting through a plethora of sensory data and memories to render the version of reality that you perceive. It's like having an internal game engine that processes inputs tirelessly, filters out the irrelevant, and highlights what feels meaningful. Yet, this remarkable process raises profound questions: Are the choices that we make truly our own, or are they merely the results of an intricate rendering of inputs that are affected by the wiring of our brains, the influence of our surroundings, and a sprinkle of randomness that nudges our decisions this way or that?

As we peer into the future, the rise of AI brings these questions into sharper focus. Just as our minds render our personal realities, AI systems operate by parsing data, training models, and making decisions based upon identified patterns. In sleek, modern data centers with rows of humming servers and glowing screens, AI models crunch numbers, render predictions, and make decisions with lightning speed. While this process mirrors the dynamics of observation and response found in game rendering and quantum mechanics, there's a crucial question: Can AI ever truly perceive and understand the world as we do, in all its complexity and nuance, or will it always be confined within the boundaries of its models and (mere) statistical data patterns?

Whether it's the vivid terrains of a video game, a quantum particle's precise position, or the thoughts and decisions that flicker through our minds, each instance is part of a grand dance of observation and realization – a continuous rendering of reality. In this light, everything that we know and experience seems to spring into being through the lens of our observation that our unique perspectives influence and define. It's a powerful thread that weaves together some of the most compelling fields of human inquiry, from game design and quantum mechanics to the mysteries of consciousness, the puzzle of free will, and the frontiers of AI.

Rendering Games

Shall we play a game?

Sitting on the edge of your favorite chair, game controller in hand, you dive headfirst into a sprawling universe, a digital expanse where anything is possible. Today's online multiplayer games have evolved far beyond mere entertainment – they're bustling, virtual ecosystems where players aren't just gamers, but explorers, strategists, and even storytellers in their own right. These platforms, like Fortnite and Valorant, are more than just digital playgrounds – they're modern frontiers where individuals can test the limits of their abilities, experiment with limitless possibilities, and engage in dynamic, often unpredictable encounters with others.

Take Fortnite, for instance. What began as a battle royale has transformed into something much larger – a multifaceted cultural phenomenon where gaming collides with social networking and live entertainment. It's a bit like a cosmic city that never sleeps, where avatars gather to compete, but also to experience events that transcend the game itself. For example, millions of players from all corners of the globe gathered in-game for a virtual concert by Travis Scott. It wasn't just a performance; it was a groundbreaking event that blurred the lines between the virtual and the real, and turned a simple game into a worldwide spectacle. Then there's Valorant, which has ignited a global passion for tactical team-based competition that has propelled e-sports to a prominence that rivals traditional sports. While reflexes and aim are important, in these arenas it's about strategy, collaboration, and the thrill of outsmarting your opponents – a digital chessboard on which every move counts.

These games aren't designed simply for passing time; they engage our minds in ways that mirror humanity's great scientific endeavors. Just as astronomers use simulations to unravel the mysteries of the cosmos, Fortnite and Valorant players immerse themselves in complex, high-stakes scenarios where every decision can spell triumph or defeat. Picture yourself coordinating a squad as a Fortnite storm closes in, each choice critical, or navigating the intricate tactics of Valorant, where deploying an agent's abilities at the perfect moment can change the course of the game. It's akin to a well-coordinated space mission, where teamwork, foresight, and precision are key. Like scientists exploring the secrets of the universe, players here are adapting, learning, and constantly strategizing, honing skills that resonate far beyond the screen.

What makes these games so compelling is their ability to tap into our deepest desires: to explore; to overcome, and to understand. In these virtual spaces, you can be anyone – hero, tactician, explorer – navigating unknown territories and tackling challenges that push you to the edge. It's this quest for discovery and mastery that motivates both gaming and the human fascination with the cosmos. Games like Fortnite and Valorant aren't just about the thrill of the moment; they are microcosms of the human experience that reflect our innate drive to connect, compete, and create.

Fortnite

Fortnite, developed by Epic Games and launched in 2017, didn't just enter the gaming scene – it exploded onto it and reshaped the landscape with a vibrant blend of battle royale chaos and whimsical, cartoonish charm that made it stand out from everything else in the genre. When the game begins, you're dropped from a flying bus onto a sprawling, colorful island with 99 other players, each one eager to be the last person standing in this adrenaline-fueled game of survival. The island is no static backdrop; as the match progresses, the map begins to shrink, pushing everyone into ever-closer quarters and making every decision more crucial. Scavenging for weapons, tools, and resources becomes a mad dash, and each encounter is a life-or-death showdown. However, Fortnite's true secret sauce is its unique building mechanic. In the blink of an eye, players can conjure up walls, ramps, and entire fortresses from thin air and craft cover or gain the high ground with the ease of a wizard with a magic wand. This layer of strategic building adds a thrilling twist that turns every firefight into a dynamic battle of wits and reflexes. Whether you're flying solo, teaming up in duos, or joining forces in squads, the ultimate goal remains: outlast everyone else.

Fortnite's origins trace back to a cooperative mode called Fortnite: Save the World, in which players banded together to fend off waves of encroaching zombie-like creatures while they defended their bases – a concept with its own appeal. However, it was the battle royale mode that truly captured the gaming world's attention, catapulted Fortnite into a global phenomenon, and left Save the World in its shadow. Yet, Fortnite is more than just a frantic scramble for survival; it's a constantly evolving experience spiced up with an array of in-game events that defy a shooter's usual boundaries. One moment, you're caught in a fierce firefight; the next, you're grooving to a live concert by the likes of Travis Scott or Ariana Grande, or even battling side-by-side with superheroes from the Marvel universe. These ever-changing events keep the game fresh, ensure that no two seasons are ever alike, and lure players back for more, season after season.

Central to Fortnite's appeal is its extensive roster of skins – characters that players can collect and wear, each of which adds a personal flair to their in-game persona. Whether you're charging into battle as a slick cyber-ninja, a towering banana, or even donning the cape of Batman, these skins are purely cosmetic, yet they infuse every encounter with personality and flair. Unlike some shooters who mix in AI opponents, Fortnite's battle royale is a pure test of human skill – every opponent is a real person, which makes each match a fresh challenge against players from around the globe. To help new players ease into the fray, Fortnite has introduced AI bots recently, which offer a gentler on-ramp to the intense competition and makes the game more accessible to newcomers.

At its core, Fortnite is a game of strategy that demands quick thinking and adaptability. Mastering the art of building is as crucial as aiming straight – knowing when to throw up a defensive wall or a sky-high ramp can be the difference between victory and a swift return to the lobby. Shooting the fastest is important, but you also need to think the smartest, read the terrain, and adapt to the ever-shifting conditions – whether that's the encroaching storm circle or the sudden appearance of a powerful new weapon that could turn the tide. Fortnite's success lies not only in its engaging mechanics, but in its relentless reinvention of itself. It's a shooter, a builder, and a living, breathing world all rolled into one – a place that feels at once familiar yet endlessly surprising.

Valorant

Valorant, which Riot Games developed and released in 2020, burst onto the competitive gaming scene, where it skillfully blended the precise gunplay of classic shooters with the strategic complexity of unique characters' abilities. Set on a near-future Earth, Valorant's world is one where agents from diverse backgrounds and countries bring more than just guns and grenades to the battlefield; they wield a distinct array of powers that can change the course of any encounter. It's a high-stakes game of chess in which every piece has its own magical twist: an agent who can summon towering walls of flame to block enemy sightlines, another who sends out drones to scout for hidden threats, or a character who lays down toxic clouds that block key choke points. At its heart, Valorant pits two teams of five against each other in a tense, strategic showdown – one side defends key sites while the other tries to plant a bomb, known in-game as the spike. The setup may echo the tactical intensity of classic titles like Counter-Strike, but Valorant's true charm lies in its agents' abilities, which turn every match into a complex mind game of bluffs, ambushes, and clever plays.

Riot Games envisioned Valorant as more than just another shooter; they intended to create a competitive experience that emphasized tactical depth, razor-sharp gunplay, and, crucially, fairness. With that vision, they crafted a robust anti-cheat system from the ground up, dedicated to catching cheaters and ensuring that every match is a true test of skill and strategy. In Valorant, you'll never face off against AI opponents in competitive matches – every round, every duel is against real players, which sharpens the edge of every decision and every shot. For those who want to hone their skills, training modes with AI bots provide a stress-free environment to practice aim, refine tactics, and master the nuances of each agent's abilities without the pressure of live competition.

Success in Valorant hinges on a careful blend of economic management – deciding when to splurge on high-powered weapons and abilities versus saving for future rounds – and tactical map control.

As your team inches through a tight corridor, nerves taut as you cover every angle, you're fully aware that the opposing team might be lying in wait, ready to unleash a perfectly timed flashbang or smoke screen. In Valorant, every decision carries weight, from when to buy a top-tier rifle or save for a critical round, to when to push aggressively or hold back. The economy system introduces a layer of strategy where teams must balance risk and reward, which makes each match a dynamic and unpredictable experience in which one clever move or mistimed purchase can entirely shift the game's momentum.

Valorant's appeal lies in its high skill ceiling, but also in its demand for seamless teamwork and constant communication. A well-coordinated team can synchronize their abilities to devastating effect by blinding opponents with a flashbang before they burst through a smokescreen for a surprise assault. This mix of tactical planning, sharp reflexes, and the ability to read and counter the enemy's moves makes Valorant a favorite among competitive players. It rewards those who can think several steps ahead, predict the enemy's moves, and set the perfect trap.

While Valorant thrives on precise gunplay and tactical maneuvers, Fortnite offers a different flavor of competitive gaming, one in which the freedom to build and destroy shapes a vibrant, ever-evolving battlefield. In Fortnite, while you're fighting, you're also creating, turning the world into your playground with every ramp, wall, and fort that you build. Meanwhile, Valorant's world is one of precision and planning, where every bullet counts and every move is a calculated risk. Whether you're scaling a makeshift tower in Fortnite or clinching a match point in Valorant with a perfectly placed headshot, these games immerse players in unique, dynamic arenas that push skill, strategy, and teamwork to their limits and offer endless possibilities in the world of competitive multiplayer gaming.

Game Engines

In the realm of digital entertainment, creating games with a game engine is like setting off on a grand space expedition with a state-of-the-art spacecraft fully equipped with the latest technology, rather than trying to piece together a rickety vessel on a deserted island. Game engines like Unity and Unreal come pre-loaded with advanced modules for rendering graphics, simulating physics, managing sound, and handling input, all of which allow developers to channel their creative energies into the artistry of game design rather than become bogged down in the technical minutiae of building these foundational systems from scratch. This approach not only reduces development time, but also speeds the entire production process that enables ideas to be transformed from mere concepts into fully realized virtual worlds.

One of the advantages of using a game engine is its capacity to be deployed across platforms. Games developed with engines like Unity or Unreal can transfer from PCs to consoles to mobile devices and thereby widen their audience with minimal extra effort. Modern game engines are replete with the latest rendering technologies that offer developers the tools to craft stunning graphics, lifelike lighting, and intricate physics simulations that bring digital worlds to life – achievements that would be daunting to gain from the ground up without these sophisticated systems.

Game engines like Unreal and Unity are vibrant ecosystems brimming with communities, support networks, and resources. They are updated continuously with the latest tools, performance enhancements, and support for emerging hardware that ensure that developers remain at the forefront of technology without having to overhaul their entire architecture with every leap forward. Built-in optimization features, such as advanced memory management and efficient resource loading, keep games running smoothly across a spectrum of devices, from high-end gaming rigs to everyday smartphones that ensure a consistent experience for all players.

For example, consider Fortnite, the global phenomenon powered by Epic Games' Unreal Engine. Unreal Engine's capabilities lie in its ability to render vast, visually spectacular environments while it manages the fast-paced, real-time mechanics that are the essence of Fortnite's allure. This engine empowers Fortnite's developers to innovate continuously and add fresh features, events, and visual updates that keep the game perpetually engaging for millions of players worldwide.

Similarly, Valorant, developed by Riot Games, harnesses Unreal Engine, but with custom modifications tailored specifically for a competitive shooter's tactical demands. Riot's meticulous tweaks are laser-focused on optimizing performance, minimizing latency, and maximizing frame rates – crucial elements in a game where every millisecond counts. Valorant's developers navigate the technical challenges of latency and synchronization, and refine their engine to deliver beautiful visuals with the pinpoint responsiveness and fairness required by competitive play's high-stakes nature.

At their core, game engines are the technological foundations of the immersive, interactive worlds that define games like Fortnite and Valorant. They are far more than simple toolkits; they are the engines that fuel the imagination, the powerful machinery that allows developers to bring their wildest visions to life to ensure that every movement, interaction, and visual detail is rendered with precision and delivered to players around the globe.

GPUs

Game engines are designed to run at their best with the latest graphics cards, and in the world of gaming, the graphics card – or GPU (Graphics Processing Unit) – is the true star of the show. The GPU transforms streams of complex digital data into vibrant, dynamic visuals that light up our screens. At the heart of a graphics card, the GPU is an incredibly efficient processor designed specifically to manage the intense parallel processing demands of rendering graphics. While a computer's central processing unit (CPU) serves as a versatile jack-of-all-trades, the GPU is a specialist that excels in the repetitive and high-speed tasks of manipulating pixels and generating images. This makes it indispensable for gaming, but also for video editing, scientific simulations, and even the intricate calculations involved in training AI models.

However, a GPU does more than just supercharge visual performance; it also assumes a substantial portion of the computational workload, which frees the CPU to manage other tasks and ensures that the entire system runs smoothly and efficiently. Modern GPUs, like those produced by NVIDIA and AMD, are equipped with hundreds or even thousands of cores, each of which works simultaneously to process the intricate details of shading, textures, and lighting effects in real time. Like a finely-tuned orchestra, in which each instrument plays its part to create a harmonious symphony, a GPU performs millions of calculations simultaneously to bring digital landscapes to life. This capability allows developers to create immersive, lifelike visuals that elevate digital experiences to new heights. Technologies like ray tracing, which mimics the behavior of light in the real world, bring an unprecedented level of realism to digital environments that transforms them from mere simulations into scenes that feel almost tangible. A GPU works tirelessly behind the scenes to turn lines of code into the dazzling visuals that define our modern digital world.

The partnership between game engines and GPUs is a powerful collaboration that extends the boundaries of what's possible, turning flat, uninspired digital landscapes into vibrant, fully immersive environments that blur the line between reality and imagination. Sheer computational power is important, but it's also about delivering finely-tuned efficiency. Game engines are optimized meticulously to harness GPUs' unique strengths, such as their unparalleled ability to perform vast numbers of calculations concurrently, which results in smoother gameplay and more detailed, captivating experiences. Innovations like NVIDIA's DLSS (Deep Learning Super Sampling) take this collaboration a step further by using AI to upscale lower-resolution images and thus deliver crisp, high-resolution visuals without sacrificing performance. This cutting-edge technology allows game developers to achieve stunning graphics and high frame rates that draw players deeper into the heart of their virtual worlds.

The ongoing collaboration between game engines and GPUs exemplifies the harmony of technology working in concert; as GPU architecture evolves, so too do game engines' capabilities. This determines digital landscapes' evolution much like the continual advancements in astronomical instruments propel our exploration of the cosmos. Game developers depend upon the symbiotic relationship between game engines and GPUs to push the envelope of digital artistry and craft experiences that are as mesmerizing as a glimpse into the depths of space.

Unreal Engine

Let's take a closer look at Unreal Engine, the brainchild of Tim Sweeney at Epic Games – a technological marvel that has revolutionized digital creation since its inception. It's a game engine designed with the precision and focus of a seasoned astronomer that meticulously calculates what needs to be seen and what can be left in the shadows. This technique, known as frustum culling, is similar to gazing through a telescope: your view is limited to a narrow slice of the sky that reveals only a few stars or galaxies at a time, while the rest remains hidden. Similarly, Unreal Engine constantly assesses the player's position and perspective to determine which objects, characters, and environments fall within the player's field of view, or frustum. By rendering only these elements, the engine conserves precious computing power and memory, ensuring that it can maintain performance and visual fidelity without overwhelming the hardware. If a game tried to render every object in the scene, even those hidden behind walls or too far away to see – it would be like trying to capture the entire night sky in one sweeping view, a task impossible for any telescope or GPU to manage efficiently. To refine this further, Unreal employs occlusion culling, which ensures that objects blocked by others are skipped even within the visible frustum, which thus optimizes the balance between visual splendor and system performance.

Unreal Engine is celebrated as one of the most advanced and versatile game engines in existence, renowned for its highly sophisticated graphics capabilities that allow developers to craft stunning and immersive digital worlds. At its core, Unreal excels in high-quality real-time rendering and uses sophisticated lighting and shading systems that bring virtual environments to life with breathtaking realism.

One of the most groundbreaking innovations within Unreal Engine is Nanite, a virtualized micro polygon geometry system that represents a seismic shift in rendering technology. Nanite allows developers to work with incredibly detailed 3D models composed of billions of polygons – something that traditionally would strain any system to its limits and force compromises in either detail or performance. Like trying to sculpt a galaxy with just a chisel, the sheer complexity makes it impossible without simplification. Yet Nanite automates this process, streaming and rendering dynamically only the exact details visible to the player at any moment. This technology allows artists to import assets from high-resolution sources like CAD data or 3D sculpting software directly so that they can bypass the cumbersome manual task of creating less detailed versions for different viewing distances. As a result, Nanite enables creators to build vast, intricate worlds that retain their visual fidelity regardless of how closely they are scrutinized, all the while ensuring that their performance remains smooth and responsive.

Beyond its visual prowess, Unreal Engine is a versatile powerhouse packed with tools for animation, physics, and AI. It's a go-to for game development, but it has also become a staple in such industries as film, architecture, and virtual reality. Its Blueprint visual scripting system democratizes development further by allowing creators to design complex game mechanics and interactions without writing a single line of code. This feature empowers both newcomers and seasoned developers to prototype and refine their projects rapidly and extend creative boundaries without the usual technical obstacles. Unreal Engine is a gateway to limitless possibilities and enables developers to bring their wildest visions to life with a blend of precision, power, and creativity that mirrors the relentless curiosity of exploring the cosmos.

Rendering AI Opponents

In the vast and ever-expanding universe of gaming, AI players and non-player characters (NPCs) are the digital stars, planets, and galaxies that bring virtual worlds to life. These AI-driven entities are carefully crafted to simulate intelligent behavior and thus create immersive experiences that dynamically respond to every move that the player makes. Think of stepping into a bustling virtual city where characters aren't just decorative figures – they're part of a living, breathing ecosystem. Whether they serve as loyal allies, cunning foes, or background characters that populate the game's landscape, these AI-driven beings are powered by sophisticated algorithms that dictate their movements, decisions, and interactions. Drawing from advanced AI techniques, such as pathfinding algorithms for navigating complex terrains, decision trees for making real-time choices, and even machine learning to adapt to evolving player strategies, these characters do more than just follow scripts – they actively engage with the world around them.

The sophistication of AI in games can range from simple, repetitive actions to highly complex systems capable of learning and evolving in real time. AI bots often step in to maintain the game's pace by filling gaps left by absent human players and ensuring a consistent level of challenge and interaction. However, these AI bots aren't mere stand-ins – they are designed with care to make the game world feel vibrant and reactive. An AI enemy in a first-person shooter doesn't simply charge blindly into battle. Instead, it takes cover, flanks you, or even coordinates with other AI units and displays a level of tactical thinking that transforms it from a simple obstacle into a worthy adversary. This complexity elevates AI players and NPCs beyond mere scripted roles – they become essential components of the game's narrative fabric who enrich the tactical environment and heighten the player's immersion.

This dynamic interplay between the player and AI mirrors the relationship between observer and observed in physics and astronomy. Just as gravity, magnetic fields, and cosmic interactions influence celestial bodies, a network of complex programming and adaptive algorithms govern AI characters and make the worlds that they inhabit feel as vibrant and unpredictable as the cosmos itself. Like when you're navigating a virtual landscape and every character reacts to your presence, or engaging in a strategic duel with an AI opponent that learns from your moves and adjusts its tactics accordingly.

In this sense, AI players and NPCs in modern gaming transcend their status as mere digital constructs. They help define spaces that feel as alive and dynamic as those in the real world. They invite players to step into a realm where every interaction builds upon the last and thereby craft a narrative that unfolds with the complexity and depth of the universe itself.

Rendering Observation

As you step into a vibrant, sprawling video game world, every move and every glance feels like opening a new chapter in an unfamiliar storybook. Here, the player's perspective functions like a powerful lens that brings the virtual environment into sharp, vivid focus, reminiscent of the quantum principle that particles exist in a state of uncertainty until observed. In game development, this technique is known as "lazy loading", "on-demand rendering", or "culling" – a clever strategy that ensures that the game world isn't fully rendered all at once.

Rendering on-the-fly isn't merely an intelligent way to conserve GPU resources; it's a delicate dance between the player's curiosity and the game's technological boundaries, crafting an experience that feels both expansive and fluid, intricate yet efficient. The player's actions bring the game world into existence, piece by piece, like navigating a dense forest where each tree and hidden path reveals itself gradually as you move forward, or spotting a distant castle that fully resolves into detail only when you approach. This ongoing interplay between the seen and unseen deepens the sense of immersion and makes the player feel like an integral part of the world's unfolding narrative.

The game engine tirelessly recalculates and redraws the environment as the player's perspective shifts. Each movement becomes a cue for the game to bring new elements into focus – characters, objects, or entire landscapes – while areas outside the player's view remain in a state of potential, ready to spring into existence as soon as the player's attention turns their way. By concentrating resources on what truly matters – the elements within the player's direct line of sight – the game engine constructs a world that feels boundless and lifelike without overwhelming the system's capabilities.

Ultimately, this interaction between player and game creates a deeply personal and uniquely responsive world. Whether uncovering hidden secrets in a sprawling cityscape or marveling at the natural beauty of a virtual landscape, the player's observational choices breathe life into the game and highlight the observer's profound role in defining the nature of their digital adventure.

As we shall see, game engines, quantum physics, life, consciousness, and artificial intelligence all share a unifying principle: the necessity of efficient rendering. Whether it's a video game world being drawn frame by frame, the probabilistic collapse of quantum states, the metabolic economy of living systems, the mind's real-time simulation of reality, or AI's streamlined decision-making processes – each operates under the constraint of limited resources and the demand for coherent output. These systems must selectively focus, compress, and synthesize vast amounts of information to create a stable, intelligible experience. Efficiency of rendering isn't just a technical requirement – it's the hidden architecture that allows complexity to function, perception to emerge, and meaning to arise.

Rendering the Quantum World

Classical physics

Classical (or naïve) physics is our instinctive understanding of how the world works – a mental toolkit filled with gut-level rules and expectations that we begin to build in early childhood. From playing catch to pouring milk, we gather practical, hands-on experience that influences our common sense about the physical world. It's the simple ideas like "heavy things fall faster" or "a ball rolls until it runs out of energy" that guide us through everyday tasks. These notions help us predict that a thrown ball will trace a neat arc through the air or that a moving object will eventually come to a stop unless something keeps it going. This intuitive grasp of physics is invaluable in daily life, as it allows us to stack books without them toppling or judge an approaching car's speed. Yet, despite its usefulness, naïve physics often clashes with the precise laws of the universe, as Newtonian mechanics or the mind-bending principles of quantum theory dictate.

Consider the common misconception that a rolling ball slows down because it "runs out of energy". In truth, friction and air resistance are doing the work – forces that aren't immediately obvious, but are crucial to the underlying mechanics. Similarly, Galileo famously debunked the classic belief that heavier objects fall faster, as he demonstrated at the Leaning Tower of Pisa that, in a vacuum, all objects fall at the same rate regardless of mass. These examples highlight the way that our everyday perceptions, while practical, often diverge from the more accurate, often counterintuitive insights derived from scientific inquiry. Naïve physics offers a visceral, first-hand grasp of the world, but falls short when faced with the formal intricacies of true physics and reminds us that the universe can surprise us by defying what seems like common sense.

The iconic tale of an apple falling on Isaac Newton's head is a perfect snapshot of our everyday understanding of gravity: what goes up must come down. This aligns neatly with our common sense that objects have weight, they fall when dropped, and that the forces acting on them appear direct and straightforward. Newton's brilliance lay in elevating this simple observation to universal principles that could explain why apples fall, why the Moon orbits Earth, and why planets trace their elegant paths through the solar system. For Newton, the apple's fall wasn't just an isolated event; it was a glimpse into the grand, unseen forces that affect the cosmos. His revelation linked the mundane with the celestial by showing that the same gravitational force that causes the apple to fall also keeps the planets in line – a profound leap that reshaped humanity's understanding of the universe.

Newtonian physics provides a framework of predictability and order that resonates deeply with our intuitive sense of how things work. It depicts the universe as a vast, predictable machine where everything operates according to well-defined laws, from an apple's fall to a rocket's trajectory. Newton's laws of motion and universal gravitation describe a cosmos in which space and time serve as the unchanging backdrop for a predictable, deterministic sequence of events. This classical view has proven extraordinarily effective in explaining a wide range of phenomena, from planetary orbits to the mechanics of bridges and the flight of rockets, and has painted the universe as a clockwork system where, given sufficient knowledge, the future is entirely predictable.

Yet, as deeply ingrained as Newtonian physics is in our understanding of the world, it isn't the final word. On the grandest scales of the cosmos and the tiniest scales of atoms, nature reveals complexities that Newton's laws cannot fully explain. Einstein's theory of relativity introduced a universe in which space and time bend and warp, and quantum mechanics unveiled a subatomic world of probabilities and uncertainties that defy classical intuition. These groundbreaking discoveries invite us to explore beyond Newton's clockwork cosmos and peel back additional layers of reality.

Emergence

As it happens, classical Newtonian physics emerges from a deeper, stranger set of principles known as quantum mechanics. Emergence is one of those wondrous phenomena where the universe reveals its most magical trick: simplicity that gives rise to breathtaking complexity, often without any grand blueprint or central conductor. As you stand in a field at dusk and watch a flock of starlings perform their mesmerizing aerial ballet, hundreds of birds twist and turn in perfect synchrony, creating swirling, fluid shapes that ripple across the sky. It's as if they've rehearsed this performance countless times, but in reality, there's no leader directing the dance. Each starling follows just a few simple rules: stay close to your neighbors, avoid collisions, and align your direction with those around you. From these basic principles, a coordinated and captivating display emerges, far greater than the sum of its parts. This is emergence in action – a beautiful reminder that complexity can effortlessly spring from simplicity.

Emergence is everywhere that you look in nature. Take an ant colony, for instance. On their own, ants are small, seemingly insignificant creatures with limited cognitive abilities, yet together, they construct intricate nests, forage efficiently, and mount formidable defenses. How do they achieve this? Each ant follows simple cues: when one finds food, it leaves a pheromone trail for others to follow. There's no overarching plan or leader guiding their actions, yet from these modest interactions, a highly organized society emerges. It's as if nature has an innate talent for creating order from chaos and conducting symphonies without a conductor.

The cosmos itself is a grand showcase of emergence on a staggering scale. Consider galaxies – those majestic spirals of stars, gas, and dark matter. These vast, beautiful structures aren't planned centrally; they emerge from the gravitational interplay of countless stars and cosmic debris, each pulling and pushing on one another and gradually sculpting the graceful arms and glowing clusters that we see through our telescopes. It's a cosmic choreography where the invisible hand of gravity brings order to the chaotic dance of celestial bodies and creates the stunning galactic tapestries that adorn the universe.

Looking deeper still, each car on the road is itself a marvel of emergence. A car is a collection of meticulously engineered parts – steel, rubber, glass – each designed for a specific role, but at a more fundamental level, these parts are made of atoms and molecules, the building blocks of all matter. It's as if every layer of the universe is another stage of emergence: from the basic elements to the complexity of machinery, to the fluid dynamics of traffic, and extending all the way to the vast structures of the cosmos.

A short history of quantum mechanics

At its deepest level, the universe is more than a collection of atoms; it's a complex tapestry woven from even more fundamental particles – protons, neutrons, and electrons – held together by forces that obey the enigmatic laws of quantum mechanics. This microscopic dance is far removed from the predictable world of classical physics and delves into the heart of quantum theory, a journey that began at the dawn of the 20th century with pioneers like Max Planck and Albert Einstein, who sought to explain phenomena that defied conventional understanding.

Prior to 1900, the so-called "black body problem" was one of those delightful puzzles that threw the brightest minds of physics for a loop. Imagine standing by a cozy fireplace; you feel the warmth on your skin and notice the embers' vibrant glow. This warmth is energy radiating from the hot coals, just as stars and planets radiate energy into space. Scientists at the time tried to explain how objects like these radiated energy across different wavelengths – essentially, how their heat translated into light – but their equations kept falling apart. Classical physics predicted that as objects became hotter, they should emit more and more light at shorter wavelengths, which leads to something absurd: infinite amounts of energy at ultraviolet wavelengths (and a painful sunburn!). This wild prediction became known as the "ultraviolet catastrophe" – theoretical physics had gone off the metaphorical rails. Enter Max Planck, who in 1900 proposed a radical solution: energy isn't emitted in a smooth, continuous stream, but in tiny, discrete packets called "quanta". This insight not only solved the black body problem, but also laid the foundation for quantum mechanics, which changed our understanding of how the universe works forever.

Einstein took Planck's groundbreaking idea further in 1905 when he explained the photoelectric effect, where light striking a metal surface ejects electrons, a discovery that would later earn him the Nobel Prize. He proposed that light itself is composed of these quanta, now known as photons, which revealed for the first time the bizarre dual nature of light – it behaves as both a particle and a wave. This concept was a profound challenge to everything that we thought we knew about light and laid the groundwork for a deeper exploration into the quantum realm. Then, in 1913, Niels Bohr introduced his revolutionary model of the atom, which suggested that electrons occupy quantized orbits and only jump between these orbits by emitting or absorbing energy. Bohr's model explained the spectral lines of hydrogen and suggested a universe where the rules were nothing like the classical laws that had governed our previous understanding.

The 1920s marked a quantum – no pun intended – leap forward with the advent of Werner Heisenberg's matrix mechanics and Erwin Schrödinger's wave mechanics, two mathematical frameworks that sought to capture subatomic particles' strange behavior. Schrödinger's wave equation proposed that particles could be thought of as waves of probability, a radical departure from the Newtonian idea that particles are solid objects traveling along defined paths. Complementing this, Heisenberg's uncertainty principle stated that it's impossible to know both a particle's exact position and momentum simultaneously, which introduced a fundamental limit to our knowledge – a revelation that struck at the heart of classical determinism and suggested that the universe was not just odd at its core, but inherently uncertain.

By 1927, the Copenhagen interpretation, influenced chiefly by Niels Bohr and Heisenberg, posited that quantum particles do not possess definite properties until they are observed, which suggested that observation plays a crucial role in affecting reality itself. This was a philosophical bombshell that redefined our understanding of existence and sparked debates that continue to this day. Later, Paul Dirac expanded the theory by merging quantum mechanics with relativity, while Richard Feynman developed quantum electrodynamics (QED), a theory that explains how light and matter interact with stunning precision and solidified quantum mechanics as the foundation of our understanding of the microcosm.

One of the most peculiar and still contentious phenomena that emerged from quantum theory is entanglement, first brought into focus by Einstein, Podolsky, and Rosen in 1935 as a challenge to the theory's completeness. Entanglement describes the way that particles, once linked, remain interconnected so that the state of one instantaneously affects the other, regardless of distance – a phenomenon that Einstein famously derided as "spooky action at a distance." Despite his skepticism, entanglement has been verified experimentally and now underpins sophisticated technologies such as quantum computing and quantum cryptography, where information can be processed and transmitted in ways that make the most advanced classical technologies seem rudimentary.

By the end of the 20th century, quantum mechanics had established itself as one of the most successful scientific theories ever conceived, underpinning everything from the chemistry of atoms and molecules to the function of semiconductors in electronics, and the lasers that scan our groceries. Yet, for all its successes, quantum mechanics continues to challenge and perplex us by defying our everyday intuitions with its insistence that particles can exist in multiple states simultaneously and that outcomes are probabilistic rather than deterministic. The reality at the quantum level differs fundamentally from the world that we experience daily: In this strange realm, as we shall see, particles can tunnel through barriers, teleport information across space, and exist in superpositions that defy the neat, orderly rules of classical physics.

Quantum mechanics and game engines

As you walk through the landscapes in the video game world, the game engine works furiously behind the scenes (generating lots of heat!), until you decide to turn back or venture farther ahead.

Remarkably, the universe behaves in a strikingly similar way. Quantum mechanics suggests that reality itself may not bother to "render" every detail of existence until we observe it. Particles and systems remain in a haze of probabilistic possibilities and exist in multiple potential states simultaneously, much like a game engine stores data for a forest that it hasn't displayed yet. The universe collapses these probabilities into a definitive outcome only when observed – when we "look" at reality. Behind the scenes of both game engines and quantum mechanics lies an intricate dance of processes, algorithms, and rules designed to balance efficiency with the complexity of existence.

At the heart of this orchestration in games is the event queue, a tool that sequences and synchronizes everything from a character's movement to the timing of a dramatic explosion. It's akin to the way that quantum systems prioritize states for measurement. Just as an event queue resolves ambiguity by processing one event at a time, quantum mechanics collapses a particle's wavefunction into a single observable state. Both systems simplify chaotic possibilities into defined outcomes by clearing a path through the fog of uncertainty.

State machines, another hallmark of game engine design, regulate the way that objects transition from one state to another – whether a door swings open, a character shifts from running to crouching, or an enemy changes from passive to aggressive. This mirrors quantum coherence, in which particles exist in a delicate superposition of states until interactions dictate their evolution. Yet, as in the quantum world, interference from external forces can disrupt these transitions. In quantum terms, this is decoherence – the collapse of superposition into a single reality. In game engines, it's the moment when uncertainty gives way to a definitive state.

Event observers play a pivotal role in both systems. In a game engine, an event observer listens for events – say, a player pressing a button or an object entering a collision zone – and reacts accordingly. This mechanism is strikingly similar to quantum entanglement. In the quantum realm, measuring one particle in an entangled pair affects its twin instantaneously, regardless of the distance. In game engines, when an event occurs, all linked event observers spring into action to ensure coherence across objects, much like quantum states in a unified system are maintained.

Signals in a game broadcast events dynamically and connect objects without direct references – just as quantum entanglement allows particles to share information across space. This elegant system creates a web of interlinked events, although it's not immune to interference. A failed signal disrupts the flow, mimics the decoherence that scatters quantum states, and breaks their once-perfect connections.

Consider the hierarchy of a scene graph in a game engine, where parent-child relationships govern how objects interact. A parent object, like a moving train, propagates events to its children, such as doors and passengers. This cascading flow is reminiscent of quantum interference patterns, in which energy and states ripple across interconnected particles. However, if a break occurs in the hierarchy – say, the train derails – the system's coherence collapses, echoing quantum decoherence.

Tags categorize events in games by grouping them into coherent sets, much like quantum coherence organizes grouped particles. These tags help filter and track related interactions – whether managing enemy behavior or logging environmental changes. Misclassification or interference among tags parallels decoherence as well by disrupting the harmony of grouped quantum states.

Timers in game engines ensure that events unfold in sequence, like a ticking clock that triggers explosions or transitions. This parallels the evolution of quantum systems that are governed by probabilistic equations over time. Misaligned timers can disrupt a game's flow, much like timing errors in quantum systems lead to decoherence and scatter the delicate order of events.

Global managers oversee game-wide interactions, similar to the way that quantum fields govern particles across the cosmos. These managers coordinate coherence among distributed objects to ensure unified behavior. A failure in the manager's communication mirrors quantum decoherence, as it leads to fragmented systems and unintended outcomes.

Physics engines track such interactions as collisions in which cascading effects mirror quantum interference. When a boulder rolls down a hill and smashes into obstacles, each collision spawns new events. This is the digital twin of particles interacting and entangling upon collision. Yet, inaccuracies in these interactions scatter coherence, much like decoherence in quantum mechanics.

Logging systems in games meticulously record events and preserve their sequence and relations. This archival process mirrors the data collection in quantum experiments that captures how states evolve over time. Any gaps or errors in the log disrupt clarity, which echoes the way that decoherence obscures the relations between quantum states.

Networking systems synchronize events across players in multiplayer games, much like quantum entanglement synchronizes states across particles, even at vast distances. Latency and packet loss in networks mimic decoherence's effects by breaking the unity of shared states. Like entangled quantum systems, networks that function properly maintain coherence and unified behavior.

Dependency graphs map the way that game events rely on one another, which reflects the interconnectedness of quantum entanglement. When one event triggers another – such as flipping a switch that powers a door – the graph ensures coherence. However, a break in the dependencies mirrors decoherence by unraveling the once-interlinked system.

Finally, behavior trees in games dictate AI decision-making, which mirrors quantum systems that are transitioning between probabilities. Each tree branch represents a possible action, much like interference patterns guide quantum states into observable outcomes. Disruptions in a behavior tree's logic reflect decoherence and break linked actions' coherence.

In both game engines and quantum mechanics, the universe – or a simulation of it – is based upon a delicate balance of rules and relations, where coherence reigns until interference tips the scales. By linking the digital and the quantum, we glimpse the profound similarities in the way that systems – real and virtual – render the worlds that we see and experience.

Let's take a step back now and explore the various aspects of quantum mechanics in greater detail.

Schrödinger's wave equation

Erwin Schrödinger's wave equation, which he formulated in 1926, stands as a cornerstone of quantum mechanics and marks a dramatic shift from classical physics' clockwork precision into a realm of uncertainty and probabilities. Consider that Newton's laws of motion, those steadfast rules that guide planets in their orbits and projectiles along predictable paths are rewritten for a world in which particles behave more like ripples on a pond than solid marbles rolling across a table. Schrödinger's equation is the key to navigating this subatomic landscape, where particles exhibit both wave-like and particle-like characteristics that defy the straightforward logic that governs our everyday experiences. At its heart lies the wave function, represented by the Greek letter psi (Ψ), a kind of cosmic map not of certainties, but of probabilities – a mathematical expression that doesn't pinpoint a particle's exact location, but rather where it's likely to be found if you were to look.

Imagine a map that doesn't mark a single point, but spreads possibilities across a vast spectrum. This is the essence of the wave function: it's like a weather forecast that doesn't predict just sunshine or rain, but lays out a cloud of probabilities for every possible outcome. The true power of Schrödinger's equation derives from the square of the wave function, which gives us the probability density, essentially the likelihood of finding the particle in any given spot. It's like searching for a needle in a haystack, except that this needle exists in multiple places simultaneously, each with a different likelihood of being discovered. This is a radical departure from the neat, deterministic paths laid out by Newton's laws, where every particle has a specific position and velocity and follows a precise trajectory like a well-aimed arrow. Instead, Schrödinger's work reveals that, at the quantum level, particles exist in a superposition of states, a haze of potential positions that collapse into a definite outcome only when observed.

This collapse of the wave function suggests that observation is not a passive act, but an active force that affects the very fabric of reality. In Schrödinger's quantum world, the act of measuring a particle doesn't simply reveal its state – it influences it and causes the particle to "choose" a specific position from a cloud of possibilities. It's as if reality itself is in a state of flux, a fluid mosaic of potentialities that only solidifies when we interact with it. This probabilistic nature of quantum mechanics challenges the classical view of a world governed by immutable laws and predictable outcomes, and hints instead that the universe is interconnected and responsive to the observer's actions. In this quantum realm, certainty yields to possibility, and every act of observation transforms the state of the observed by collapsing the wave function's broad spectrum into a single, measurable point.

As we saw with rendering in a game engine, the act of looking isn't just the act of seeing; it's the act of influencing the very reality that we observe, a reminder that in the deepest layers of existence, the simple act of observation is intertwined with the fabric of the cosmos itself.

Wave-particle duality

Wave-particle duality is a mind-bending and paradoxical concept in quantum mechanics that reveals that particles like electrons and photons can exhibit both wave-like and particle-like behaviors depending upon how we observe them. Thomas Young was the first to dramatically demonstrate this concept in the early 19th century through his famous double-slit experiment. Young shone light onto a screen through two closely-spaced slits and predicted that if light were purely a stream of particles, it would produce two distinct lines that corresponded to the slits. Instead, the light created a pattern of alternating light and dark bands – a classic signature of wave interference. It was as if the light waves were overlapping, amplifying in some places and canceling out in others, which produced a bright and dark striped pattern on the screen. This experiment suggested that light, which was traditionally thought of as a particle, also behaved as a wave.

Physicists revisited Young's setup on the 20th century with a curious twist: they fired individual photons through the slits one at a time. Their expectation was that each photon would travel through one slit or the other, and create two distinct marks on the screen like a marble tossed through a doorway. Instead, even when photons were sent through one at a time, the interference pattern reappeared, slowly but surely, as though each photon passed through both slits simultaneously and interfered with itself. This baffling result shattered the classical expectation that particles have a definite position and trajectory and revealed that at the quantum level, reality does not conform to the predictable rules of Newtonian physics. Instead, it appeared that particles could exist in a state of superposition and take every possible path until observed.

Moreover, this quantum quirkiness wasn't limited to light. When physicists conducted the same double-slit experiment with electrons – long considered the quintessential particles – they too produced an interference pattern. It appeared that rather than selecting one slit to pass through, these tiny electrons behaved like waves that passed through both slits at once and created an interference pattern on the screen behind. It was like watching electrons play a cosmic game of hide-and-seek, flitting through both slits simultaneously and interacting with their own probability waves. This strange behavior defied the Newtonian view of particles moving along clear, predictable paths dictated by forces. In the quantum realm, entities like electrons exist as a spread of potential locations, represented by waves, and only "collapse" into a specific spot when measured.

This phenomenon underscores the profound and perplexing nature of wave-particle duality. Rather than behaving like predictable marbles, electrons act like waves and (appear to) take every possible route simultaneously. Yet, the moment we use a detector to observe which slit the electron passes through, this wave-like behavior vanishes; the interference pattern disappears, and each particle "chooses" a slit and collapses from a superposition of possibilities into a single, definite outcome thereby.

This strange shift highlights the idea that reality isn't a solid, fixed landscape at the quantum level, but a fluid realm of probabilities that crystallize into certainty only when observed. It tells us that on the smallest scales, the universe is not as rigid and predictable as Newton described. Instead, it is governed by probabilities and possibilities, where particles can exist in multiple states simultaneously until observed. Experiments like the double-slit offered profound revelations that challenged our deepest intuitions about the nature of reality. They revealed that what we once saw as fundamental distinctions between waves and particles are, in fact, two aspects of the same quantum phenomenon. This duality blurs the lines between the tangible and the abstract and paints a picture of reality far more mysterious and fluid than the classical mechanics of Newton ever dared to imagine.

Superposition

Perhaps the most famous illustration of superposition comes from Schrödinger's cat, a thought experiment that Schrödinger conceived in 1935 to highlight quantum mechanics' mind-boggling implications. A cat is sealed within a box with a radioactive atom, a Geiger counter, a vial of poison, and a hammer ready to shatter the vial. It's like a twisted game of chance: if the Geiger counter detects radiation – triggered by the atom's random decay – the hammer is released, smashing the vial and releasing the poison, which kills the cat. If no radiation is detected, the cat remains alive. According to quantum mechanics, until the box is opened and the system observed, the atom exists in a superposition of decayed and undecayed states, and this uncertainty extends to the entire setup. This means that, within the confines of the box, the cat is simultaneously alive and dead, a bizarre scenario that defies common sense and underscores the strange nature of quantum superposition.

In the familiar world of Newtonian physics, objects have clear, definite properties – positions, velocities, states – that exist independently of observation. A ball is either on the ground or in the air, moving or still, and its behavior can be calculated with absolute precision based upon initial conditions and forces. There's no room for ambiguity; everything runs like clockwork and every action has a clear and predictable outcome. However, superposition stands this intuitive understanding on its head. At the quantum scale, particles exist in a hazy cloud of potential states with no single reality until an observation compels one of those possibilities to become the definitive outcome. This probabilistic nature of quantum mechanics reveals a universe far less deterministic than Newton envisioned, where observation doesn't merely measure reality, but actively affects it.

One of the enduring puzzles of quantum mechanics is why and how this "collapse" from superposition to a definite state occurs when measured – a conundrum that left Schrödinger and generations of physicists after him grappling to determine the true nature of reality. Does this collapse require a conscious observer? Until the moment of measurement, a quantum system exists as a blend of all possible states, much like a cacophony of overlapping notes waiting for a listener to zero in on one and turn it into the melody. Superposition isn't simply a theoretical oddity; it's the cornerstone of quantum computing, where bits of information – qubits – can exist in multiple states at once, which allows them to perform complex calculations that classical computers could never match. However, this phenomenon also forces us to reconsider the fabric of existence itself, as it suggests that at its most fundamental level, reality is not as solid and definite as our everyday experiences – or Newtonian physics – might suggest. Instead, it's a dynamic, fluid tapestry of possibilities, being constantly shaped and reshaped by the act of observation.

Entanglement

Quantum entanglement is another baffling and counterintuitive phenomenon in quantum mechanics that reveals a hidden web of connections that defy our classical understanding of the way that the universe operates. Two particles, such as electrons or photons, become intertwined so completely that the state of one instantaneously determines the other's state, regardless of the distance that separates them. This "spooky" idea unnerved Einstein because it seemed to violate the bedrock principles of locality and causality that are fundamental to classical physics. In the Newtonian view, objects are independent entities that only interact through local forces such as gravity or electromagnetism, and information cannot travel faster than the speed of light. Yet entanglement suggests that on the quantum level, particles can be linked in a way that transcends space and time and hints at a profound interconnectedness woven into the very fabric of reality.

Entanglement's true nature was put to the test in the 1980s by physicist Alain Aspect and his team, who conducted a series of landmark experiments that would alter our perception of the quantum world forever. Aspect's experiments were designed to test Bell's theorem, which physicist John Bell formulated in the 1960s. Bell's theorem argued that no theory based upon local hidden variables – essentially the classical idea that particles have predetermined states – could fully reproduce the predictions of quantum mechanics for entangled particles. This raised a fundamental question: Do particles have predetermined states that we reveal through measurement, or do quantum mechanics' strange, non-local rules genuinely govern their behavior?

To explore this mystery, Aspect and his team focused on entangled pairs of photons – particles of light – whose polarizations were related in ways that defied classical explanation. They built a sophisticated apparatus to measure each photon's polarization after the pair had been emitted and traveled apart, which thus ensured that the measurements were made independently. To eliminate the possibility that the settings could influence the results through conventional, local means, Aspect ingeniously designed the experiments with rapidly changing detector settings and chose what to measure only after the photons were already on their way. This rapid switching was crucial, as it ruled out any possibility that hidden variables could "know" the settings in advance and adjust the photons' behavior accordingly.

The results were nothing short of revolutionary. Aspect found that when one photon's state was measured, its entangled partner's state was determined instantly, even when separated by distances too vast for any signal to travel between them at light speed. This instantaneous effect defied the classical notion of locality, which insists that only their immediate surroundings can influence objects. Instead, Aspect's experiments provided compelling evidence that entangled particles share a connection that transcends the constraints of space and time, which suggested a universe where information can leap across vast distances, as if ignoring the very fabric of spacetime itself.

Is there a way to send a stream of entangled photons across the galaxy with the goal to communicate by measuring some while leaving others untouched – a sort of cosmic Morse code whispered among the stars? Unfortunately, quantum mechanics doesn't allow for "peek" measurements where you can check a particle's state without causing its collapse. The moment you interact with a photon, even gently, it collapses from its state of possibilities into a single outcome. Without disturbing it, there's no way to observe it; any measurement forces the particle to reveal itself and disrupts its quantum superposition.

Aspect's experiments were pivotal, as they provided concrete experimental validation for quantum mechanics' predictions and demonstrated the quantum world's fundamentally non-local nature. His work challenged the classical view of a predictable, clockwork universe and laid the groundwork for future technologies that harness the peculiar properties of quantum mechanics, such as quantum computing and cryptography.

Quantum Tunneling

Quantum tunneling is an intriguing phenomenon in the quantum realm in which particles defy the classical rules that seem so fundamental to our everyday understanding. If a ball rolls up a hill in the familiar Newtonian world, it needs sufficient energy to crest the top and descend the other side. However, in the quantum universe, particles like electrons play by different rules. Rather than climbing over the hill, they seem to disappear from one side and reappear on the other, as if by magic. This is quantum tunneling – a process that defies the classical notion that an object must have sufficient energy to overcome an obstacle. It's as if the particle has found a hidden shortcut and slipped through barriers that would be insurmountable in the macroscopic world.

The first significant clue to this quantum sleight of hand emerged in the 1920s when physicist George Gamow applied it to explain alpha decay, in which alpha particles escape from an atom's nucleus. These particles don't have the classical energy needed to break free from the strong nuclear forces that bind them, yet they manage to "tunnel" through the barrier as though it doesn't exist. Gamow's insight was revolutionary, as it challenged the very foundations of classical physics, and has since been confirmed by countless experiments. Quantum tunneling isn't simply an abstract idea confined to the theoretical realm or exotic lab conditions – it has practical applications that have transformed technology and even the way that we explore the world.

One of the most remarkable uses of quantum tunneling is in the scanning tunneling microscope (STM), a device that allows scientists to visualize surfaces at the atomic level by exploiting electrons' tunneling across a minuscule gap between a sharp tip and the sample. It's like using a needle to "feel" the terrain of a surface by delicately tracing the dips and rises at the atomic scale to reveal the invisible landscape of atoms and molecules. The STM doesn't just look at atoms; it uses quantum tunneling to "feel" them, similar to mapping a mountain range with the light touch of a feather. This tool has opened a new window into the microscopic world and allowed us to see and even manipulate individual atoms with unprecedented precision.

Quantum tunneling has parallels in the natural world also, e.g., by echoing the way lightning behaves. When a bolt of lightning strikes, it doesn't travel in a straight line from the cloud to the ground. Instead, it explores multiple potential paths through branching, invisible tendrils called "stepped leaders", much like quantum particles exploring different possibilities in superposition. Just as lightning searches for the path of least resistance, quantum particles exist in a cloud of probabilities, "feeling out" the chance to be on the other side of a barrier. When conditions align perfectly, the particle tunnels through, bypassing the need for the classical energy altogether.

From a classical standpoint, quantum tunneling appears to be nonsensical; Newtonian mechanics tells us that an object must have sufficient energy to overcome any barrier. However, in the quantum world, particles aren't confined to acting as solid objects – they exist as probability waves, which allows them to explore possibilities that defy classical logic. Here, probabilities replace certainty, and the seemingly impossible becomes routine.

The Uncertainty Principle & Non-determinism

The Uncertainty Principle that Werner Heisenberg introduced in 1927 is one of the most thought-provoking and fundamental concepts in quantum mechanics, as it challenges our classical notions of measurement and precision. If you take a snapshot of a bustling city at night, the lights blur, cars streak by, and the more you try to focus on one detail, the more others dissolve into the darkness. This beautifully captures the essence of Heisenberg's principle: At the quantum level, you can never know a particle's position and momentum with perfect accuracy at the same time. The more clearly you try to capture one, the fuzzier the other becomes. Heisenberg illustrated this through a thought experiment that involved a hypothetical microscope, and envisioned the attempt to pinpoint an electron's exact position. The act of measurement itself would disturb the electron's momentum and make it impossible to determine both properties simultaneously. This isn't attributable to any flaw in our tools or techniques – it's a fundamental characteristic of the quantum world, a stark departure from the precision we expect in the orderly Newtonian universe.

Quantum mechanics introduces a level of non-determinism that turns our understanding of the universe on its head and reveals that at the deepest levels, events are governed not by certainty but by probabilities. In the Newtonian world of classical physics, everything appears to work like the synchronized gears of a giant clock; if you know the system's current state, you can precisely predict its future state. However, in the quantum realm, the universe resembles a foggy crossroads where, regardless of how carefully you prepare or calculate, you can never be entirely sure which path will be taken until it's already behind you. This is the crux of quantum uncertainty: particles exist in a state of potential and do not adopt a definite state until observed or interacting with something else.

As mentioned, one of the most striking manifestations of this non-determinism is seen in radioactive decay, a phenomenon that captivated early 20th-century scientists like Ernest Rutherford. Radioactive decay occurs when unstable atomic nuclei release energy in the form of particles, and while we can calculate the probability that an atom will decay within a specific timeframe – much like predicting the odds of rolling a six on a die – the exact moment of decay for any single atom is fundamentally unpredictable. This randomness isn't a consequence of insufficient measurement or missing data; it's woven into the fabric of nature itself. Each atom seems to make a spontaneous decision about when to release its energy, which defies the deterministic laws that comfortably describe the larger, everyday world.

The double-slit experiment was also mentioned, in which individual particles like electrons or photons created an interference pattern when allowed to pass through two slits. The pattern overall can be predicted only after many particles have completed their journey, but any individual particle's path – whether it goes through the left slit, the right slit, or even through both slits in a mysterious superposition – remains fundamentally indeterminate until it hits the screen. Even then, probability, not any deterministic rule, governs each particle's exact impact point.

In Newton's classical worldview, the universe was similar to a vast, finely-tuned machine, in which every action led to a predictable reaction and the causes and effects were chained together in a precise, unbroken sequence. Theoretically, if you could measure every particle's position and velocity, you could predict the unfolding of events past, present, and future with unerring accuracy. However, quantum mechanics disrupts this orderly vision and replaces the solid gears of determinism with the fluid probabilities of an improvisational performance. Rather than a universe marching to a predetermined script, we find ourselves in a realm where particles move in waves of probability that allow for multiple, simultaneous potential outcomes that collapse into a single reality only when observed. The outcome realized is often determined by a roll of the quantum dice.

Quantum computing

The race to develop quantum computers is one of the most exhilarating frontiers in modern science that echoes the fervor and ambition of the 20th century space race. It's a global contest that pits some of the brightest minds and most advanced technologies against each other, all striving to create machines that could revolutionize our understanding of computation. Unlike classical computers, which process information in bits that exist as either 0s or 1s, quantum computers exploit the peculiar principles of quantum mechanics and use qubits that can exist as 0, 1, or both simultaneously through superposition. This ability to be in multiple states at once allows quantum computers to process vast amounts of information in parallel, and address problems that would be insurmountable for even the most powerful classical machines.

A group of scientists huddles around a device that looks like a cross between a futuristic chandelier and a steampunk contraption – a quantum computer. These aren't your standard desktops; they operate at temperatures colder than outer space, just fractions of a degree above absolute zero. This extreme cold is crucial because qubits, the building blocks of quantum computing, are incredibly delicate. They can lose their quantum state at the slightest disturbance – a phenomenon known as quantum decoherence, mentioned earlier – which renders their calculations useless. To prevent this, quantum computers are housed in cryogenic chambers that protect them from heat, vibrations, and electromagnetic interference and thereby create an environment as serene as a cosmic winter.

Building a quantum computer is like constructing a house of cards in the middle of a windstorm, with qubits made from exotic materials such as superconducting circuits, which allow electricity to flow without resistance, or trapped ions suspended in electromagnetic fields. A single particle, floating perfectly still in a vacuum, is manipulated into different states by meticulously calibrated lasers. This is a trapped ion qubit, a marvel of quantum engineering. Other qubits are made from photons that zip through optical circuits guided by beamsplitters and mirrors in a precise dance of quantum choreography.

Quantum computers' true potential lies in their ability to perform numerous calculations simultaneously thanks to superposition and entanglement, where qubits become so intertwined that the state of one instantaneously influences the state of another, regardless of the distance between them. This interconnectedness allows quantum computers to explore many possible solutions at once, which dramatically speeds up tasks that would take classical computers millennia. Think of navigating a vast maze with countless branching paths. A classical computer would methodically explore each route one by one, and check every possibility until it found the exit. However, a quantum computer can traverse all paths simultaneously and find the exit almost instantaneously. This extraordinary capability allows quantum computers to revolutionize such fields as cryptography, optimization, material science, and complex simulations – challenges that are currently beyond the reach of conventional computing.

However, the journey into the quantum realm is just beginning. Today's quantum computers are more like early prototypes, capable of only limited calculations and prone to errors from decoherence and other forms of quantum noise. The goal is to build larger, more robust quantum processors in which thousands, or even millions, of qubits work together. These future quantum machines could transform fields such as drug discovery by allowing scientists to simulate molecular interactions with unmatched accuracy or break existing encryption standards and render today's unbreakable codes obsolete.

As the competition increases, tech giants like Google and IBM, together with ambitious startups, are vying to build the most powerful quantum computer. In 2019, Google announced a significant milestone with its quantum processor, Sycamore, and claimed that it had achieved "quantum supremacy" by completing a specific calculation in 200 seconds that would have taken the world's fastest classical supercomputer an estimated 10,000 years. This achievement was a landmark akin to the Wright brothers' first flight – proof that quantum computing wasn't simply theoretical, but a tangible reality, albeit one still in its infancy and applicable only to a narrow set of problems.

Quantum computing brings the principle of efficient rendering into sharp focus, revealing how the universe itself may function like a vast, optimized computation. Unlike classical computers that process bits one at a time, quantum systems "compute" by evolving entire probability spaces in parallel, resolving only the meaningful result at the moment of measurement. This mirrors the universe's own strategy: maintaining a probabilistic cloud of possibilities until resolution is absolutely necessary. Quantum computation doesn't just use the rules of quantum physics – it exemplifies them.

Rendering Life

Does observation create reality?

Unlike the comforting predictability of the Newtonian world, in which objects exist independently of our gaze and move along predefined paths, quantum mechanics reveals a far more fluid and interconnected universe, one deeply entwined with the observer. As discussed above, in this bewildering quantum landscape, particles acquire definite properties or locations only when they are measured, suggesting that observation is not merely a passive act, but one that actively brings physical phenomena into being. This Observer Effect challenges the classical notion of an objective reality that exists independently of us, and hints at a universe in which observation and reality are inextricably related, and each influences the other in ways that defy traditional logic.

John Archibald Wheeler, a visionary physicist known for his imaginative leaps in understanding, took this idea even further with his concept of the "participatory universe". Wheeler proposed that the act of observing the universe contributes to its creation, and encapsulated this idea in the phrase "It from Bit." He suggested that physical reality arises from informational processes, and posited that the universe becomes real through acts of measurement and observation. Wheeler's "delayed-choice experiment" is a striking illustration of this notion, as it presents a quantum twist that defies everyday logic. Choices made in the present can influence events that occurred in the past, thus blurring the conventional lines between cause and effect. For example, consider starlight that has traveled across the cosmos for billions of years; according to Wheeler, the moment that it reaches your eye, it retroactively determines whether it traveled as a wave or a particle. It's as if the universe holds its breath, awaiting our observation to decide how it should behave, which makes the observer an integral part of the unfolding cosmic drama.

Eugene Wigner, another giant in quantum mechanics, explored the intricate relation between consciousness and reality and proposed that consciousness plays a crucial role in collapsing the wave function. Wigner went further and suggested that it's not just the act of measurement that causes this collapse, but the conscious awareness of the measurement. He vividly illustrated this idea through a thought experiment known as "Wigner's Friend." Imagine you have a friend inside a laboratory who measures a quantum system while an observer outside the lab considers the entire setup – including the friend – as part of a larger superposition. A paradox arises when the perspectives of the friend and the external observer conflict, which highlights the perplexing implications of observation and consciousness in the quantum realm.

Wigner's theories suggest that consciousness may not be merely a byproduct of brain activity, but could be a fundamental aspect of the universe that exerts influence at the quantum level. While speculative, this notion challenges the conventional view of the universe as an observer-independent entity and suggests instead that consciousness and reality are deeply intertwined. Both Wheeler and Wigner's groundbreaking perspectives invite us to reimagine the universe not as a static stage where events unfold, but as a dynamic, participatory arena where the observer plays a crucial role in affecting what is real. These ideas stretch the boundaries of both physics and philosophy and raise profound questions about the nature of existence and the role that the conscious observer plays in the cosmos.

Although still the subject of intense debate and far from universally accepted, the concepts of a participatory universe and the role that consciousness plays in quantum mechanics offer a tantalizing glimpse into a reality where the familiar rules of cause and effect dissolve and are replaced by a universe that responds to the act of observation itself. From this perspective, the line between observer and observed isn't just blurred – it may not exist at all in the traditional sense. It's as if the universe, in some profound way, is aware that we are watching and is making us participants in its ongoing creation.

What is life?

In 1944, Schrödinger published his groundbreaking book *What is Life?*, a visionary work that sought to disentangle the complex threads that connect quantum physics and biology. At a time when the mysteries of life seemed nearly insurmountable, Schrödinger boldly applied the principles of physics to biological questions, which set the stage for the emergence of modern genetics and molecular biology. One of the most remarkable ideas that he introduced was the concept of an "aperiodic crystal" – a stable yet intricately structured molecule able to store genetic information. This prescient notion remarkably foreshadowed the discovery of DNA as the carrier of genetic material, a revelation that James Watson and Francis Crick would provide just a decade later. With direct inspiration from Schrödinger's insights, they would go on to unravel DNA's double helix structure. Schrödinger's vision of life as an ordered system governed by quantum mechanics challenged the scientific paradigms of his time by suggesting that genetic replication's stability and precision might hinge on the strange, unpredictable laws of the quantum realm.

Schrödinger also delved into the paradoxical relation between life and entropy, and tackled one of the great conundrums of biology thereby: How do living organisms maintain their intricate order in a universe dominated by the second law of thermodynamics, which dictates that entropy, or disorder, must always increase? He proposed the concept of "negative entropy", or "negentropy", to describe the way that living organisms sustain their organized complexity. He suggested that life extracts order from its surroundings by importing energy, whether through sunlight or food, and exporting entropy back into the environment. This energy exchange allows organisms to maintain their highly ordered states and fend off the natural drift toward chaos. Schrödinger's idea reframed life as a relentless battle against entropy sustained by the continuous inflow of energy that supports the construction and maintenance of the complex structures necessary for survival.

What is Life? is more than just a scientific treatise; it's a philosophical exploration that investigates the nature of consciousness, free will, and the essence of being alive. Schrödinger's interdisciplinary approach bridged the gap between physics and biology and inspired a generation of scientists to think beyond the conventional boundaries of their fields. By combining the quantum mechanics of the microscopic world with the macroscopic phenomena of life, Schrödinger invited us to reconsider the fundamental nature of existence itself.

His insights continue to resonate and urge us to explore the intricate interplay between the physical laws that govern the universe and the mysterious, self-sustaining phenomenon that we call life. Through *What is Life?* Schrödinger opened new windows onto the profound connections between the orderly universe described by physics and the dynamic, complex tapestry of living organisms. His work pushed us to reflect on the deeper questions of the way that life emerges from the inanimate and challenged us to think about the universe as not simply a collection of particles and forces, but as a place where the laws of physics and the wonder of life intersect in unexpected and awe-inspiring ways.

How did life first evolve?

The story of life's origins takes us back to a world vastly different from the one that we know today – the Hadean or early Archean eon approximately 3.8 to 4 billion years ago. Earth was a cauldron of primordial forces, in which volcanoes erupted with molten rock and belched clouds of ash and gas into a thick, noxious atmosphere. Asteroids frequently pummeled the surface, shaking the nascent planet to its core and leaving it a hostile landscape of boiling seas and unformed landmasses. Yet, amid this seemingly inhospitable chaos, the first sparks of life began to flicker into existence, in which early bioactive chemical reactions set the stage for the emergence of life as we know it. These reactions likely took place in environments that resembled a primordial soup – shallow pools rich in minerals, scalding hydrothermal vents on the ocean floor, or even within the protective crannies of mineral-rich rocks. In these varied settings, simple organic molecules – formed from essential compounds like methane, ammonia, and water – began to interact, driven by the raw energy sources of the time: ultraviolet radiation from the Sun, lightning strikes crackling through the turbulent skies, and the geothermal heat seeping from the planet's molten interior.

The first significant glimpse into these early processes came in 1952 in Stanley Miller and Harold Urey's groundbreaking experiment. They sought to recreate the conditions of early Earth within a glass apparatus filled with water, methane, ammonia, and hydrogen designed to simulate the primordial atmosphere. By bombarding the mixture with electrical sparks to mimic lightning, they set the stage for a chemical symphony reminiscent of the planet's tumultuous beginnings. Astonishingly, after just a week, they discovered that amino acids – the fundamental building blocks of proteins – had formed within their simulated environment. This experiment was a watershed moment in the study of life's origins, as it demonstrated that life's essential ingredients could spontaneously arise under the proper conditions. Their results suggested that Earth's volatile early environment was not an obstacle, but rather a crucible for life's formation, as it provided the raw materials and energy needed for life's earliest steps.

Over the course of eons, these simple molecules engaged in a complex dance of interactions that wove together the biochemical tapestry that would eventually lead to nucleic acids, proteins, and other macromolecules crucial for life. This intricate process crossed a critical threshold and led gradually to the emergence of self-replicating entities – the precursors to the first primitive cells. These early bioactive reactions marked the dawn of biochemistry, when inert chemicals began to assemble into dynamic, living systems. It was a pivotal transition, as it transformed the planet from a barren rock into a world teeming with life. From these humble beginnings, life as we know it sprang forth and turned Earth into a vibrant realm of complexity and diversity.

The Miller-Urey experiment and subsequent discoveries have continued to support the profound idea that life's building blocks are not confined to Earth or some cosmic rarity; instead, they may emerge naturally wherever the conditions are right, potentially even on distant planets or moons. As we look out into the cosmos, we carry with us the understanding that life's essential ingredients – those first flickering sparks of existence – are not beyond the bounds of possibility. From the chaotic crucible of early Earth to the vast reaches of space, the journey of life is a testament to the power of chemistry and the resilience of nature, a journey that began in the most unlikely of circumstances, but continues to shape our world and perhaps many others beyond.

The first observer

While pinpointing the exact moment when life first influenced a quantum event is beyond our reach, it likely occurred more than 3 billion years ago during the early chapters of Earth's history. At that time, the simplest life forms, such as bacteria and archaea, were already interacting with their environment in ways that reflected the fundamental principles of quantum mechanics. A compelling scenario involves the advent of photosynthesis, a process that allowed primitive organisms to harness sunlight using bioactive molecules like chlorophyll. These molecules absorbed photons that set off quantum events that excited electrons – a phenomenon that can be viewed as an early instance of wave-function excitation at the quantum scale.

Eventually, photosynthesis transformed Earth's atmosphere because of ancient photosynthetic organisms like cyanobacteria. At the core of this process, chlorophyll molecules operate on a quantum level by capturing photons and using their energy to convert carbon dioxide and water into glucose and oxygen. This light-harvesting mechanism relies on quantum effects such as excitation energy transfer and quantum coherence, which efficiently guide energy through the chlorophyll molecules and make photosynthesis remarkably effective. These molecules' ability to funnel energy through quantum processes allowed early life to thrive in environments that would have been inhospitable otherwise.

Another early interaction in life that may have triggered wave-function excitation involves primitive biochemical reactions facilitated by enzymes. These enzymes catalyzed essential reactions that depended upon quantum effects like tunneling, in which particles defy classical physics by passing through seemingly insurmountable energy barriers. Such quantum interactions were crucial for the biochemical processes that sustained early life and marked some of the initial moments when living organisms began to influence the quantum realm. Essentially, as soon as life on Earth began to engage in chemical interactions governed by quantum mechanics – whether capturing light, transferring energy, or catalyzing reactions – these events marked the dawn of life's effect on the quantum scale.

It's important to recognize that while chlorophyll's interaction with sunlight is deeply rooted in quantum mechanics, it does not involve wave-function collapse in the traditional sense found in quantum experiments. Instead, it involves quantum excitation and energy transfer, in which the photon's energy is absorbed and efficiently channeled without a definitive collapse triggered by measurement. Quantum coherence within the photosynthetic complex allows the energy to navigate through the system and effectively "chooses" the most efficient path – an elegant use of quantum processes to determine vital biological functions.

These early uses of quantum mechanics on the part of life forms represented a significant evolutionary leap that allowed sunlight to be converted into a dependable energy source. This quantum capability fueled early photosynthetic organisms' growth and proliferation and set the stage for the Great Oxidation Event, which dramatically altered Earth's atmosphere and laid the foundation for the evolution of complex life. These interactions highlight the profound relations between the microscopic quantum world and the macroscopic evolution of life on our planet, which illustrates that even the most basic organisms were intertwined with the quantum fabric of reality from the very beginning. It is a testament to the deep and unexpected ways in which life has always been enmeshed with the fundamental forces that shape the universe by taking advantage of the subtle power of quantum mechanics to sustain and propel itself forward through time.

Quantum biology

You stand at the edge of a still, expansive lake just before dawn. The water's surface is so smooth that it mirrors the early morning sky with a nearly magical clarity and reflects every delicate cloud and the first golden rays of sunlight. In this tranquil scene, even the smallest ripple spreads gracefully and sends elegant, synchronized waves that glide across the lake without losing their form. This serene image beautifully captures the essence of quantum coherence – a key phenomenon in quantum mechanics that defies our everyday experiences. In quantum coherence, particles like electrons or photons exist in a delicate superposition by maintaining a precise relation with each other. It's as if they are synchronized dancers who move in perfect harmony regardless of the distance between them. This interconnectedness echoes through quantum entanglement, in which one particle's behavior can instantly affect another, even when they're separated by vast distances.

However, just like the still lake, quantum coherence is exquisitely fragile. Yet, despite this fragility, quantum mechanics has found ways to weave itself into the fabric of life itself. One remarkable example mentioned above is quantum tunneling, which plays a vital role in the chemistry of life. Consider testosterone, a hormone essential for muscle growth and mood regulation. For testosterone to work its magic in the body, hydrogen atoms must move between different parts of its molecular structure. As they are incredibly light, hydrogen atoms are particularly adept at quantum tunneling, which allows them to pass through energy barriers that would slow the hormone's effects otherwise. This process ensures testosterone's efficiency in promoting muscle growth. Think of testosterone molecules as couriers racing through a crowded city, finding hidden shortcuts – quantum tunnels – that let them bypass obstacles and swiftly deliver their messages. Without tunneling, these biochemical reactions would slow and the hormone wouldn't function as effectively.

This ability of quantum tunneling is a fundamental part of the way that testosterone and many other molecules operate. It's a vivid reminder of the universe's interconnectedness, where the rules of quantum mechanics ripple through to influence the biological processes that sustain life.

For instance, take the green sulfur bacteria, which thrive in the sun-starved depths of the ocean, where light is a rare and precious resource. These single-celled organisms have perfected a quantum trick that allows them to capture every stray photon of sunlight with astonishing efficiency. Imagine you're in a pitch-black cave, desperately trying to catch the faintest glimmer of light. For green sulfur bacteria, missing even one photon could mean death, but they've evolved to ensure that no photon goes to waste. Within their cells, structures called chlorosomes – tiny warehouses packed with light-harvesting molecules – work using quantum coherence. When a photon strikes, the energy doesn't follow just one path toward the reaction center where photosynthesis occurs. Thanks to quantum superposition, it explores multiple routes simultaneously, like a hiker finding every shortcut through a dense forest and instantly choosing the best way forward. This quantum efficiency has left scientists in awe, as these bacteria convert the faintest sunlight into life-sustaining energy.

Even more astonishing, nature uses quantum mechanics in other unexpected ways. For example, the European robin navigates on its epic migratory journey with the help of quantum effects. These birds travel thousands of miles with remarkable precision, even when skies are overcast and landmarks are hidden. For years, researchers puzzled over how robins chart their courses until they discovered a quantum secret within the birds' eyes. The robin's retina contains proteins called cryptochromes, which react to sunlight by creating pairs of entangled electrons. These electrons remain connected, regardless of how far apart they are, and respond to the Earth's magnetic field. This quantum process gives the robin a faint visual cue, a subtle inner compass that overlays an invisible map onto the sky and guides it on its journey with extraordinary accuracy. It's as if these birds have their own quantum GPS, powered not by technology, but by the invisible laws of the universe itself.

These incredible examples from nature not only captivate biologists, but they also inspire new technologies. Future solar panels that mimic the quantum tricks of green sulfur bacteria may capture sunlight with far more precision than anything we have today. It's a glimpse into how nature's quantum lessons may one day help solve some of humanity's most pressing energy challenges. The strange, elegant dance of quantum mechanics is everywhere – from the cells of a tiny bacterium to the grand migratory journeys of birds – which shows us that the universe operates in ways more profound and surprising than we ever imagined.

Evolution

This interplay between life and quantum mechanics is just one chapter in the grand, unending saga of evolution – the relentless process that has shaped the tapestry of life on Earth through the forces of natural selection. Evolution is a blind, but masterful, engineer that tirelessly refines every aspect of the living world through the straightforward rule of survival. Over billions of years, this process has led to the astonishing diversity of life that we see today, in which each species is finely tuned to its environment. At its core, evolution works by favoring traits that enhance an organism's ability to survive and reproduce, while it steadily weeds out the less advantageous and builds on the successes. This mechanism, which Charles Darwin famously described as "survival of the fittest", has optimized everything from the way that birds fly to how fish swim, and down to the molecular processes that sustain life.

In 1831, Darwin, a quiet naturalist who transformed our understanding of life on Earth, embarked on a five-year sea voyage aboard the HMS *Beagle* that would change the course of science forever. Darwin meticulously observed finches in the Galápagos Islands, where the subtle differences in their beak shapes caught his attention – variations that would become key to his theory of natural selection. He realized that these small changes were not random; they were adaptations determined by the finches' environment that allowed them to survive where others could not. This was the essence of "survival of the fittest" – those suited to their surroundings best thrived and passed on their traits, while those unsuited did not. Darwin's discovery was simple yet radical: species weren't immutable, but evolved over time through natural selection in which only the most adaptable survived. (Note that individuals don't change and adapt; species do over countless generations.) Years later, as he sat at his desk, piecing together these observations, he saw how variation, natural selection, and survival of the fittest were the determining forces of the evolution of life itself.

Evolution's ingenuity has crafted an array of biological wonders, like the molecular "walker" known as kinesin. These tiny proteins march along microtubules within the body's cells, carrying essential cargo such as nutrients and genetic material from one part of the cell to another. Each step they take is a testament to the precision of evolutionary design, honed through eons of incremental changes. The development of such molecular machines highlights evolution's capacity to devise highly specialized solutions to life's challenges by creating mechanisms that are not only effective, but also incredibly efficient and intricate.

This relentless process of natural selection operates with unwavering pragmatism and shapes life to meet the demands of its environment. It is not a random journey, but a methodical process in which advantageous traits are preserved, and maladaptive traits are not perpetuated. Evolution doesn't simply solve problems; it refines solutions by layering new adaptations atop old ones in a continuous innovation cycle. Whether it's enzymes' adaptation to function in extreme temperatures, the evolution of flight in birds and insects, or the development of complex social behaviors in mammals, evolution is determined by ceaseless pressure to adapt and thrive.

DNA

DNA is nothing short of a masterpiece of molecular engineering – a complex and elegant molecule that serves as the blueprint for all living organisms. Its iconic double-helix structure that James Watson and Francis Crick unveiled in 1953 is one of nature's crowning achievements, a twisting ladder of life composed of sequences of four nucleotides: adenine; thymine; cytosine, and guanine. These nucleotides are strung together in precise sequences that encode the vast genetic library passed down from parent to child through the eons. The genetic code is like a vast library of life, in which each gene is a chapter filled with instructions to build the structures of living organisms, while it also chronicles the story of evolution. This library isn't a static collection – it's an active participant that tirelessly guides the development, function, and maintenance of every living cell, from the simplest bacterium to the most complex human being.

The magic of DNA lies in its ability to replicate with astonishing fidelity during cell division, thus ensuring that each new cell receives an exact copy of this genetic blueprint. This meticulous replication process is fundamental to life's continuity, as it preserves the advantageous traits shaped by billions of years of evolution and passes them on to future generations. It's a dynamic contrast to non-living matter, such as a rock on the Moon, which remains inert and unchanged even as it holds the silent imprints of ancient cosmic events.

While a rock is a passive witness to time that never alters in response to its environment, DNA is deeply embedded in life processes and actively shapes the course of life by driving change and adaptation.

DNA's role doesn't end at replication; it's also involved in the cell's daily operations through the intricate processes of transcription and translation. During transcription, segments of DNA are copied into RNA, which then serve as a template for assembling proteins – the essential workhorses of the cell. This translation process transforms genetic instructions into physical actions and results in proteins that catalyze metabolic reactions, build cellular structures, fight infections, and enable movement. Through this seamless flow of information, DNA transforms genetic potential into tangible traits, whether it's the vibrant colors of a butterfly's wings, the towering height of a redwood tree, or the neural complexity of the human brain.

In addition to its role in daily cellular function, DNA is also the engine of evolution. Mutations – random changes in the DNA sequence – introduce new genetic variations that can be beneficial, harmful, or neutral. Natural selection acts on these variations and favors those that increase an organism's likelihood to survive and achieve reproductive success. Over time, this process refines the genetic code by embedding lessons learned from past challenges and triumphs into future generations' DNA. This continuous adaptation to an ever-changing environment highlights DNA's unparalleled ability not only to store and replicate information, but also to evolve, bridging the past, present, and future of life on Earth.

Intelligent Design (ID)

Words like "design" and "purpose" have long incited debate in discussions of evolution and created a delicate line between science and philosophy. The concept of Intelligent Design (ID) weaves a rich tapestry that spans from ancient philosophy through religious thought to modern scientific discourse. In the Book of Genesis, the Bible narrates God's creation of the world in six days, which culminated with mankind and all animals fully formed. Greek philosophers like Plato meticulously imagined a divine craftsman who shaped the cosmos with precision and intent, which echoed the idea of a purposeful creator. This notion found renewed vigor in the 18th century with William Paley, a clergyman who famously compared the complexity of life to a watch – so intricate that it must have been crafted by an intelligent watchmaker. Paley's analogy struck a chord, as it suggested that the elaborate systems observed in nature could not exist without a designer's guiding hand.

As we've seen, this comforting narrative faced a seismic challenge in 1859 when Charles Darwin published *On the Origin of Species*. Darwin's theory of natural selection constituted a radical departure from the notion of purposeful design, as it proposed that life's diversity and complexity arose from a gradual, unguided process attributable to environmental pressures and random mutations. This was a revolutionary and, for many, unsettling idea, yet it offered a robust, naturalistic explanation that redefined our understanding of the natural world by emphasizing that life's intricate forms could emerge without a designer's blueprint.

Despite Darwin's profound influence, the allure of purposeful design persisted and resurfaced in the late 20th century under the banner of ID. Figures like Phillip E. Johnson championed this movement and argued that natural processes alone could not explain life's complexity, particularly in structures deemed "irreducibly complex" like the bacterial flagellum. Michael Behe's 1996 book *Darwin's Black Box* popularized this view further and asserted that specific biological systems are so complex that they must have been designed, as they could not function without all of their parts in place. However, the mainstream scientific community largely dismissed ID as repackaged creationism, arguing that it lacked empirical support and failed to provide a testable scientific framework.

The debate took an unexpected turn with the advent of modern bioengineering, which blurred the lines between nature's randomness and purposeful design.

Human designers

In recent decades, scientists have harnessed their understanding of DNA to create new life forms in the laboratory. Craig Venter's work stands as a striking example: in 2010, Venter and his team created the first synthetic organism, *Mycoplasma mycoides JCVI-syn1.0*, by assembling synthetic DNA from scratch. This bacterium, which was crafted in a lab rather than through natural reproduction, marked a profound milestone – humans, themselves products of evolution – had become the designers of life.

Venter's achievement extends beyond the technical marvel of creating new life forms; it underscores a deeper implication that evolution, in a sense, has produced beings capable of design. Through its relentless process of mutation, selection, and adaptation, evolution has not only generated life, but has also crafted species that can reflect, innovate, and manipulate the fundamental building blocks of biology. This challenges traditional notions of ID and suggests that the capacity to design is an emergent property of evolution itself.

Consider CRISPR-Cas9, a revolutionary gene-editing tool derived from a bacterial immune system. Originally a natural defense mechanism, scientists have adapted CRISPR to edit genes with unprecedented precision that thereby allow deliberate changes to DNA that once seemed like science fiction. This technology, a product of evolutionary processes itself, allows us to alter various organisms' genetic code. It is as if evolution, in its blind, yet boundless creativity, has handed us the keys to a new realm of biological design.

These scientific advancements reveal a profound truth: The intelligence that we apply in human-directed design is a natural outgrowth of evolution's creativity. The "blind watchmaker" that Richard Dawkins described has, through natural selection, crafted organisms able to wield the tools of creation – a poetic irony that blurs the lines between randomness and purpose. By fostering adaptability and innovation, evolution has led to a stage at which life reacts not only to environmental challenges, but can also actively shape its destiny.

The story of ID, from ancient philosophical musings to sophisticated genetic engineering, reflects an ongoing debate about purpose and process in the natural world. Traditional ID posits an external designer, but the capabilities now at our fingertips suggest that design is a continuum within evolution's scope. This evolving understanding illustrates that the quest for meaning in life's complexity is not a fixed pursuit, but a dynamic exploration that expands as we delve deeper into the very fabric of existence. Far from being a mere mechanism, evolution emerges as a grand narrative of creativity, adaptation, and the relentless pursuit of life's potential – one that constantly finds and exploits even the slightest advantage, as nature's ceaseless drive for survival has played out over billions of years.

Maxwell's Demon

The second law of thermodynamics, which dictates that all systems evolve naturally toward disorder and increased entropy, stands as one of the immutable pillars of physics. This principle governs everything from the cooling of hot objects and the diffusion of gases to the ultimate fate of the universe itself, all of which are moving inexorably toward a state of maximum disorder. However, in 1867, the brilliant Scottish physicist James Clerk Maxwell dared to challenge this seemingly inviolable rule with a thought experiment that has remained one of the most provocative in the history of science. Maxwell imagined a tiny, intelligent being – whimsically dubbed a "demon" – that was able to sort gas molecules between two chambers based upon their speed and energy. Maxwell's demon was not merely a playful jab at the second law; it was a radical proposition that opened new avenues of thought about the nature of information and its profound role in the physical universe.

Maxwell's demon, which possesses extraordinary intelligence and agility, is a diligent sentinel perched between the two chambers. This demon meticulously observes each passing molecule's speed, and deftly opens and closes a tiny gate between the chambers at precisely the right moments. By selectively allowing the faster molecules to cluster in one chamber and the slower ones in the other, the demon seemingly creates a temperature gradient (difference) without any energy input – a feat that, if achievable, would upend the second law of thermodynamics, which insists that in any closed system, entropy must always increase. This thought experiment invites us to imagine a world in which this fundamental rule could be bent, and in which order could be snatched from the relentless advance of chaos with a mere flick of the demon's gate.

Maxwell's demon is not simply a whimsical construct from the Victorian imagination; it is a profound challenge to our understanding of the laws that govern the universe. By suggesting that knowledge about the states of molecules could be harnessed to decrease entropy, Maxwell was proposing a revolutionary idea: That information itself might have a tangible, physical effect on the world. This wasn't just an intellectual puzzle; it laid the groundwork for later developments in statistical mechanics, information theory, and even quantum mechanics. Maxwell's demon became a conceptual bridge between the abstract world of information and the concrete laws of thermodynamics that demonstrated the way that knowing or measuring can manipulate the physical universe. Far from merely being a theoretical curiosity, it serves as a powerful metaphor for the limits of human understanding and the surprising ways that information can alter the course of physical processes.

The thought of Maxwell's demon sparked intense debate and captivated the minds of physicists and philosophers alike. Its paradoxical implications challenged conventional wisdom and provoked endless discussions about the subtleties of measurement, observation, and information's fundamental role in the physical world.

It wasn't until the mid-20th century, with the groundbreaking work of scientists like Leo Szilard and Rolf Landauer, that the deeper implications of the demon's actions began to be understood. Szilard demonstrated that the demon's act of measurement itself required energy and thus ensured that even Maxwell's demon could not escape the fundamental laws of physics. Later, Landauer formalized this understanding by showing that information processing is inherently related to energy dissipation, which preserved the second law of thermodynamics' inviolability. Landauer's principle established that erasing information from a system increases entropy, and thus demonstrated that the demon's clever trick ultimately fails to cheat the laws of thermodynamics.

To visualize this in a modern context, consider Bitcoin mining, which generates significant amounts of heat. Bitcoin mining involves solving complex mathematical problems using powerful computers, known as mining rigs, which typically consist of high-end GPUs. These devices perform vast numbers of calculations per second and consume a large amount of electrical power in the process. The heat generated is a practical illustration of the way that processing information – even in the digital age – still adheres to the fundamental laws of thermodynamics.

Maxwell's demon has since transcended its origins and influences fields as diverse as statistical mechanics and quantum computing. By daring to question entropy's inevitability, Maxwell's thought experiment has fueled explorations into the fundamental limits of computation, energy efficiency, and the nature of time and information in the universe. Maxwell's demon reminds us that the universe's complexities often lie in the intricate dance between energy, information, and the inexorable march of time, and it continues to inspire us to explore the boundaries of what is possible in the ever-expanding cosmos of scientific thought.

Life is information

Life is a dance between information and entropy. This sets living organisms apart from inanimate matter. The universe is a grand stage on which everything, from the smallest grain of sand to the farthest star, is subject to entropy's relentless pull. Amid this cosmic tendency, life emerges as the ultimate rebel, a system that not only stores and processes information, but also tirelessly battles against this inevitable march toward chaos. From the precise architecture of DNA spiraling within a cell to the delicate interconnections of an entire ecosystem, life stands as a relentless architect of order, defiantly pushing back against the universe's inclination toward entropy.

DNA, a chemical structure that acts as a sophisticated repository of information, is at the heart of this defiance. Think of DNA as a meticulous script, a set of encoded instructions that direct every facet of an organism's development, behavior, and reproduction. It's as if each strand of DNA carries a chronicle of evolutionary triumphs that meticulously records the adaptations that have allowed life to thrive across countless generations. This genetic code is far more than a mere sequence of molecules – it's a living library of survival strategies that orchestrate the assembly of proteins, the growth of cells, and the replication of life itself. Applying information theory – a mathematical framework that quantifies data storage, transmission, and processing – offers a powerful tool to understand how life optimizes the flow of genetic information. This optimization is a mechanism for survival; it's a battle against the forces of chaos, against molecular missteps that could unravel the intricate tapestry of life.

Life is a dynamic equilibrium of information and energy management, where the ability to store, replicate, and use information is intimately associated with the capacity to channel energy efficiently. Unlike inanimate objects that passively succumb to the universe's entropic drift, living organisms act as localized pockets of order that wield information as a blueprint and energy as a tool to construct and sustain complexity. By perpetually counteracting entropy, life carves out havens of structure and purpose in an otherwise chaotic cosmos, and thus showcases the remarkable interplay between the laws of physics and the tenacity of biological systems.

Life's unyielding determination to harness energy and process information transforms the raw materials of existence into an organized, living mosaic that is in stark contrast to the entropic pull that governs the remainder of the cosmos. It's a vivid reminder of the relentless push toward complexity and order that defines living systems – a testament to the power of information and energy in the grand tapestry of existence.

In the end, quantum rendering offers a striking reinterpretation of the age-old question: what is life? From Wigner's idea that consciousness may play a role in collapsing the quantum wavefunction, to Maxwell's demon selectively sorting particles and challenging our notions of entropy, we see hints that life is not a passive passenger in the universe, but an active participant in its unfolding. Perhaps observation is not merely a passive act but the final step in a cosmic rendering pipeline, where the probabilistic haze of quantum mechanics crystallizes into reality through the lens of awareness. In this view, life is not separate from the quantum world – it is its most refined expression, the renderer and the rendered, shaping and shaped by the very information it brings into being.

Rendering the Mind

Idealism

Bishop George Berkeley, an 18th-century philosopher, proposed some of the most provocative ideas about reality and perception that continue to challenge our understanding today. At the heart of Berkeley's philosophy is immaterialism, or subjective idealism, which boldly asserts that the physical world does not exist independent of our perception. To Berkeley, the only true entities are minds and their ideas – everything that we experience is a construct of our perceptions. He famously distilled this revolutionary perspective into the phrase "esse est percipi", or "to be is to be perceived."

As you walk through a dense forest where the earthy scent of pine fills the air, rays of sunlight filter through the canopy overhead. In Berkeley's view, the towering trees, the gentle rustling of leaves, and even the ground beneath your feet exist solely in your mind as you perceive them. When you close your eyes or walk away, these elements cease to exist in any tangible form. Berkeley argued that without a perceiver, objects like trees, mountains, or distant stars do not persist; they vanish from reality altogether. According to him, the so-called "external world" is merely a collection of sensory experiences – sights, sounds, smells, and textures – woven together by the perceiver's mind.

This view stood in stark contrast to the materialist perspective of the time, which posited that the physical world exists independent of human perception. Berkeley's philosophy suggested that we live not in a world of solid, enduring objects, but in a grand mental rendering in which every scene is conjured by the act of perception. This raises profound questions: If the universe is merely a patchwork of perceptions, what exists beyond what we can observe? What happens to the forest when no one is there to see it?

Berkeley's ideas were revolutionary and turned the prevailing assumptions of his era on their head. While materialists insisted on a world of matter that exists independently, Berkeley argued that reality is a mental construct – a vivid panorama of sensory experiences woven by consciousness. This perspective blurs the lines between the observer and the observed and suggests that reality is not an external, unchanging entity, but something intrinsically linked to perception.

Berkeley's ideas resonate with modern scientific concepts, such as the paradox of Schrödinger's cat in quantum mechanics, which mirrors Berkeley's assertion that without perception, reality remains undefined and unresolved. In the quantum realm, a particle's state resolves into a definite outcome only upon observation, which echoes the observer's central role in defining reality. Similarly, Berkeley's philosophy invites us to reconsider our understanding of existence, not as a passive backdrop, but as something that consciousness actively influences.

This suggests that we, as perceivers, are not mere spectators in the universe, but are co-creators of the reality that we experience. By placing the mind at the heart of existence, Berkeley didn't simply challenge the materialist worldview; he opened the door to deeper inquiries into the mysteries of perception, existence, and the cosmos.

Do we live in a simulation?

The Matrix, which was released in 1999, is a groundbreaking science fiction film that shatters our understanding of reality and compels us to question whether the world we perceive is genuine or merely an intricate illusion. The film plunges us into a dystopian future when intelligent machines have enslaved humanity within a meticulously crafted simulated reality. This virtual world, known as the Matrix, mirrors every nuance of real life so convincingly that most people remain blissfully unaware of their predicament. The protagonist, Neo, finds his seemingly ordinary existence upended when he discovers the shocking truth: his entire life has been an elaborate lie, part of a vast digital construct designed to keep humans docile while they are harvested as a power source for their machine overlords. Neo's journey becomes a quest not simply to fight against oppressive technology, but to unravel the nature of his reality – a cinematic echo of age-old philosophical inquiries about perception, existence, and what it means to be truly free.

The Matrix taps into enduring philosophical questions reminiscent of René Descartes' musings on perception and reality. Descartes famously speculated that our sensory experiences could be the deception of a powerful, malevolent being – a notion eerily mirrored by the machines' digital illusion in *The Matrix*. As Neo delves deeper into the truth, he confronts a classic philosophical quandary: How can we be sure that our experiences are real? This provocative question captivated audiences and sparked intellectual discussions that resonated not only with moviegoers, but also with contemporary philosophers and technologists. One of them, Nick Bostrom, expanded on these ideas in 2003 with the *Simulation Hypothesis*. Bostrom argued that if future civilizations develop advanced technology able to create highly detailed simulations of reality, it becomes statistically probable that we are living in one of these simulations rather than in "base" reality. His theory suggests that advanced beings might run numerous simulations of their ancestors, such that simulated realities vastly outnumber the singular real one. This hypothesis challenges our fundamental assumptions about existence and implies that our world could be an elaborate tapestry of data and code that blurs the lines between reality and simulation.

The allure of simulated realities extends beyond *The Matrix* into other works of fiction that explore similar themes. In *Ready Player One*, the narrative unfolds in a future when people escape the bleakness of their everyday lives by immersing themselves in the OASIS, a sprawling and immersive virtual reality universe. Unlike the Matrix, the OASIS is not a deceptive trap, but a voluntary retreat, a digital playground where individuals can become anyone and do anything. Yet, it raises profound questions about the value of experiences in simulated worlds versus those in tangible reality. Characters in *Ready Player One* grapple with the seductive nature of the virtual, as the allure of limitless possibilities in the OASIS blurs the distinction between digital contentment and real-world fulfillment. It paints a vision of the future in which the boundaries between the digital and the real become increasingly indistinct and prompts us to reconsider what it truly means to lead an authentic life.

These narratives tap into deeper philosophical debates about the nature of reality and our place within it. If our experiences, emotions, and memories can be fully replicated in a simulation, what does that imply about the authenticity of our lives? Are we merely sophisticated programs running within a cosmic supercomputer, or is there an underlying truth beyond the digital façade? *The Matrix, Ready Player One*, and Bostrom's Simulation Hypothesis compel us to explore the fabric of existence, which straddles the line between science fiction and philosophical exploration. They push us to confront the possibility that our perceived world might be more akin to a vast virtual construct than the objective reality that we have always assumed.

The Blank Slate

In the 17th century, the influential philosopher John Locke introduced a groundbreaking concept that changed our understanding of human nature and potential forever: the idea of the "blank slate", or *tabula rasa*. A newborn child's mind is an untouched canvas, a pristine surface awaiting the first brushstrokes of life's experiences. According to Locke, this mind enters the world devoid of any pre-existing knowledge or innate ideas, completely open and malleable, ready to be shaped by the environment. Every sight, sound, and sensation – the warmth of a parent's touch, the soothing rhythm of a lullaby, the vibrant hues of a blooming garden – becomes a brushstroke that gradually fills this canvas and crafts the individual's abilities, knowledge, and character over time. Locke's perspective was revolutionary; it suggested that we all begin from the same place, with equal cognitive potential, and that our experiences, education, and personal choices mold us into who we become.

This idea contrasted sharply with the prevailing beliefs of Locke's time, which often emphasized innate differences among people as the defining factors of their capabilities and destinies. Locke's theory laid the foundation for the concepts of personal responsibility and free will and proposed that if we begin as blank slates, we are not confined to predetermined paths or inherent limitations. Instead, we hold the power to make moral choices, influence our futures, and be accountable for our actions. In Locke's view, every person has the potential for growth, change, and self-determination, which highlights the transformative power of experience and the decisions that we make along our journey. It was a hopeful vision, one that suggested that through education, effort, and the right environment, anyone could achieve great things.

However, Locke's vision of the blank slate has not gone unchallenged. Modern research into cognitive development and genetics suggests that while experience plays a critical role, certain aspects of our cognition, behavior, and personality are hardwired from birth. This raises profound questions about the extent of our freedom to determine our paths. For instance, if someone displays an innate talent for music or mathematics, is it purely the result of experience and practice, or do genetic predispositions play a significant role? This tension between the blank slate and innate capacities lies at the heart of the enduring debate over nature versus nurture, free will, and moral accountability.

The debate complicates Locke's optimistic notion of equal potential, as it suggests that the influence of genetics may confer advantages or create obstacles beyond one's control. If our actions and abilities are indeed influenced by factors beyond our control, such as inherited traits, it raises difficult questions about how we understand personal responsibility and the true extent of our capacity for self-determination. Locke's idea of the blank slate remains a powerful metaphor that captures the dynamic interplay between our experiences and our inherent traits, and continues to affect discussions on human potential, equality, and the power of personal growth in the face of life's inherent complexities.

Mental concepts & representations

In a bustling city square, you're enveloped by the symphony of urban life: the hum of conversations; the rhythmic honking of cars, the clatter of footsteps on pavement. Words and concepts flow effortlessly as you navigate this familiar scene – whether it's recognizing a dog, identifying a tree, or pondering notions of justice. To most of us, this seems like a natural consequence of learning and experience. However, the influential philosopher and cognitive scientist Jerry Fodor proposed a radical twist on this understanding: he argued that every concept that you use in that moment – whether it is "dog", "tree", or "justice" – isn't something that you've pieced together from your experiences, but rather, it's something that you were born with. According to Fodor, these basic concepts are not learned or constructed; they are innate, hardwired into our brains from birth.

Fodor's theory was as provocative as it was groundbreaking, as it challenged the prevailing notion that our minds are blank slates influenced solely by experience. He suggested instead that the human mind is like a vast filing system, pre-stocked with an immense array of concepts, each ready to spring into action when triggered by the right cue from the environment. For example, when you see a dog, you don't form the concept of "dog" by mentally stitching together such attributes as fur, barking, and wagging tails. Instead, you recognize it instantly because your mind has a built-in, innate concept of "dog" that aligns seamlessly with your perception.

This bold idea stemmed from Fodor's dissatisfaction with traditional empiricist views that concepts are learned through experience or are built from basic sensory inputs. He conducted thought experiments to illustrate these theories' limitations and argued that not all concepts could be derived purely from experience. Take, for instance, the concept of a "carburetor". According to traditional views, one would need to encounter various carburetors and observe their functions to fully grasp the concept. However, Fodor pointed out that even individuals who have never seen a carburetor in action can understand the concept when it's described to them, indicating that the mind possesses an innate capacity to recognize or integrate new ideas beyond direct sensory experience.

Fodor also garnered support from the remarkable speed and accuracy with which children acquire language – far beyond what would be expected if they were learning purely through exposure and reinforcement. He cited the "poverty of the stimulus" argument, which posits that the linguistic input that children receive is too limited and imperfect to account for the complex rules of language that they grasp intuitively. For instance, children don't need to be explicitly corrected to know that "goed" is incorrect as the past tense of "go"; they naturally refine their understanding of language's nuances guided by an internal, pre-existing framework. To Fodor, this suggested that our minds come preloaded with a sophisticated system for processing concepts and language, rather than constructing them from scratch.

When you walk through a garden for the first time, you can identify flowers, trees, and effortlessly sense the concept of beauty. According to Fodor, your ability to make these distinctions doesn't come from assembling these ideas piece by piece from sensory data, but from tapping into an innate cognitive repository of concepts. In his view, the mind isn't a blank canvas waiting to be painted with experience; it's more like a library filled with books – each one a concept just waiting to be opened.

Fodor's theory has ignited debates and inspired further exploration within cognitive science. It has prompted us to rethink the extent to which our mental landscape is affected by nature versus nurture. His ideas challenged the established views of his time and suggested that much of what we consider learning is more a case of uncovering what was always there, laying dormant in the depths of our minds. Whether it's recognizing a familiar face in a crowd or grappling with abstract concepts like freedom or justice, Fodor would argue that you're not simply learning as you go – you're rediscovering elements that have been encoded within you all along, nestled in the innate architecture of human cognition.

What is redness?

Consider the cacophony of modern life: blaring honks of impatient drivers; snippets of conversation from passersby, and the rumble of a bus navigating through traffic. To our ears, these are distinct, familiar sounds, each layered over the other to form a symphony of everyday existence. Yet, in truth, our ears don't "hear" sounds in the conventional sense. Instead, they detect tiny fluctuations in air pressure – vibrations rippling through the atmosphere – and convert these mechanical waves into electrical signals. Delicate hair cells within our inner ear act as vigilant sentinels that pick up these vibrations and translate them into signals that travel along the auditory nerve to our brain. There, in the labyrinth of neurons, these raw data streams are transformed into the vivid tapestry of hearing – whether it's the melody of a favorite song, the soothing tones of a familiar voice, or the chaotic din of a busy street.

This intricate dance of perception isn't limited to hearing; it's a fundamental principle that underlies all of our senses. Consider our sense of smell: our nose doesn't directly detect "scents". Instead, it identifies chemical compounds in the air, which are then converted into signals that our brain translates into the familiar aromas of freshly brewed coffee, blooming flowers, or burning wood. This everyday act of smelling even dips into the realm of quantum mechanics. Unlike most other senses that rely on straightforward biochemical processes, olfaction may use quantum effects to discern specific scents. Recent theories suggest that our noses don't just identify the shapes of odor molecules as thought previously, but also sense their vibrational energy, which may involve the quantum tunneling phenomenon discussed earlier.

Ultimately, the external world lacks intrinsic qualities like "redness", "sound", or "scent". These are mental constructs, intricate interpretations crafted by the brain from raw sensory inputs. Picture a red leaf drifting gently from a tree on a crisp autumn day. What does it truly mean to "see" that red leaf? At its most fundamental, seeing involves light reflecting off the leaf, entering our eyes, and stimulating photoreceptor cells in the retina that respond to specific wavelengths of light – those between 620 and 750 nanometers, typically associated with the color red. Yet, the redness that we perceive isn't a property of the leaf itself; it's a vivid experience generated entirely within our minds. Our brains don't merely process the numerical data of light; they embellish it with mental concepts – colors, shapes, and emotions – that exist only in subjective experience, not in the objective, physical world.

As another example, think about the bright yellow birds, flowers, and tennis balls you see on TV. However, your TV screen consists only of red, green, and blue pixels – there is no actual yellow. Your mind perceives yellow by interpreting the combination of red and green light, thereby essentially creating the color in your imagination!

This phenomenon extends beyond individual perceptions to reflect a broader truth about human experience: much of what we perceive to be reality is not a direct representation of the external world, but rather a sophisticated construction of our brains. Our minds don't just passively receive sensory inputs; they interpret, organize, and sometimes even actively fabricate details to create a coherent and meaningful picture of the world around us. Take the richness of a sunset's colors or the vibrancy of a red rose; these aren't dictated solely by the physical properties of light frequencies, but also by how our brains contextualize and interpret these inputs with influences from factors such as surrounding light conditions, our memories, and even our emotions at the moment.

This insight is highly consistent with the philosophy of Immanuel Kant, who argued that our perception of the world is not a direct window into an objective reality, but rather a subjective construction affected by our sensory and cognitive frameworks. Kant posited that we never perceive the world in its unfiltered state; instead, we experience a version crafted by our minds, a blend of raw sensory data and the mental structures that we use to interpret it. Modern neuroscience and psychology have reinforced this view and shown that our brains are far from passive receivers; they are dynamic architects of our perceived reality that weave our sensory inputs into a personal, subjective experience of the world.

Thus, the redness of an apple, the blue of the sky, or the green of a forest is not "out there" in the external world, but exists within us, rendered by the brain's ceaseless work of translating sensory input into coherent experience. This realization challenges our conventional understanding of reality, as it reveals that much of what we consider solid and true is a sophisticated mental construction.

Constructing reality

Our brains are not passive receivers of sensory data; instead, they are active interpreters that filter, sort, and continuously integrate data to build the rich, dynamic experience of reality that we navigate daily. This ongoing mental construction involves creating and refining internal models of the world – simulations that help us track objects through space, predict changes, and understand the relations between cause and effect.

These cognitive processes' profound depth was vividly demonstrated in the work of Belgian psychologist Albert Michotte in the 1940s. Michotte conducted experiments that illuminated the degree to which our understanding of causality is innate. In a dimly-lit room, you see two colored squares – a red one and a blue one – projected on a screen. As you observe, the red square glides toward the blue one, nudges it, and the blue square darts away. Instantly, your mind perceives causation: the red square made the blue one move. Michotte's experiments showed that this perception isn't simply a learned association; it's an automatic, deeply embedded function of the brain. When the timing or movements were altered – if the squares didn't touch or if there was a delay – this sense of causality faded. His work revealed that our brains are hardwired to connect events into coherent narratives and assign cause and effect to them almost instinctively.

This cognitive instinct extends beyond simple physical interactions. It permeates our understanding of abstract concepts like goals, plans, and memories, and provides a framework that helps us make sense of our world. Consider the way that we navigate through familiar environments: As we move, our brains create mental maps that track our position and guide us from place to place. These maps are not static; they are enriched with information about the locations of objects, the timing of events, and the underlying causes of what we experience. Our working memory temporarily holds onto relevant details, our attention focuses on what matters most, and our beliefs adapt as we accumulate new information. This dynamic mental choreography allows us to adapt to our surroundings, and update our strategies and perceptions in response to new experiences.

What sets our cognition apart is our remarkable capacity for prediction. We draw on past experiences to forecast future scenarios, and refine our actions based upon what we expect might happen. This foresight extends into our social interactions, where we infer others' intentions and motivations, which helps us navigate complex social landscapes with greater ease. Language plays a pivotal role in this process, as it not only allows us to communicate thoughts and share information, but also influences the way that we conceptualize and interact with the world. Through language, we can extend the reach of our minds, articulate our ideas, and influence others in ways that transcend our immediate surroundings.

These intricate cognitive processes converge in our ability to plan and act, which is determined by a hierarchy of goals and sub-goals that guide us toward long-term aspirations, such as pursuing an education, building careers, or nurturing relationships. Our actions are fueled by internal motivators – hunger, fear, curiosity, ambition – that encourage us to engage with our environment and strive for fulfillment. As we pursue these goals, we constantly refine our behaviors in response to feedback, and thus learn from both our successes and failures. Underlying all of this are cognitive models and algorithms that allow us to juggle multiple streams of information, apply learned principles across different contexts, and approach problems with creativity and intention.

Our capacity for imagination amplifies these cognitive abilities further, which allows us to explore scenarios beyond the immediate present, replay past experiences, and test new ideas. In the theater of our minds, we generate hypotheses, envision alternative paths, and strike a balance between exploring uncharted territory and making the most of familiar ground. It's in this expansive realm of imagination that our consciousness truly flourishes and reflects the complex interplay between our mental simulations and the external world, all of which are influenced by our motivations, fears, and aspirations. Our brains' extraordinary pattern-recognition skills not only help us navigate the intricacies of everyday life, but also empower us to dream, create, and explore possibilities that extend far beyond the tangible.

The mind is a simulation

Walking through a park in autumn and you spot a lone, colorful leaf fluttering to the ground. In that instant, your mind spins a complex web of associations and emotions that may evoke a melancholy memory of change, or a fleeting thought about the impermanence of beauty. This is the mind's interpretive power at work as it transforms raw sensory data into a rich, personal tapestry of meaning. It's a skill that sets conscious beings apart from inanimate objects like rocks or trees, which, despite their existence in the physical world, remain oblivious to the layers of significance that such moments can hold.

While the physical world undeniably exists, my experience of it is entirely mediated by a mental simulation that operates according to its own internal rules. This mental construct is consistent with the sensory data, but transcends it as it weaves a complex, multidimensional representation of reality. For example, even after the red leaf has vanished from sight, my mind can still recall it vividly, dwell on its beauty, or imbue it with personal symbolism. In this way, my perception of the leaf extends far beyond mere wavelengths of light – it encompasses emotional responses, aesthetic judgments, and narrative threads that make the experience uniquely mine.

At the core of this cognitive process is a vast neural network, much like the sophisticated algorithms used in AI. Every sensory input – whether it's the sight of that leaf, the sound of footsteps on gravel, or the scent of freshly-cut grass – is mapped onto this intricate network of neurons, woven into a rich tapestry of thoughts, memories, desires, and motivations. My brain performs a kind of real-time data compression that distills the essence of each experience while it filters out extraneous details. From this complex web emerges my consciousness – a parallel universe filled with its own objects, relationships, and rules, distinct from the physical world yet intertwined with it deeply through the act of perception.

This neural symphony gives rise to the full spectrum of conscious experience – feeling the soothing warmth of the sun, the bittersweet pang of nostalgia, or the exhilaration of a long-sought triumph. My mind doesn't merely process data; it imbues them with meaning and creates a dynamic mental landscape populated by emotions, intentions, and abstract concepts such as "missed opportunities" or "hidden motives". For instance, when I see someone with a sleek new smartphone, my brain doesn't just recognize its design; it stirs a web of thoughts and feelings – perhaps a flicker of envy, a sense of aspiration, or a reminder of past desires – all woven intricately into my mental universe. My reaction isn't dictated solely by the physical object, but is affected by the rich tapestry of experiences and associations that define my perception.

Ultimately, the intricate interplay of these mental objects is what gives rise to consciousness and emotion. When I perceive the redness of the leaf or feel a surge of envy, I am engaging only with entities within my mental simulation – wait, Bishop Berkeley was right after all? – crafted by my brain's extraordinary interpretive powers. My mind doesn't simply react to the world outside; it actively constructs a personal, vibrant universe that is filled with layers of meaning and significance, each guided by the unique neural connections that shape my perspective on reality. Where is a "missed opportunity"? Only in my mind.

In this private mental universe, my senses provide the raw data, but my consciousness weaves them into the rich, subjective tapestry of my life. This intricate dance of perception, memory, and imagination allows me not just to exist, but to deeply engage with the world, and crafts a personal narrative that reaches far beyond the mere numerical data of the physical realm, which, in any case, lie unrendered unless I observe them.

In a world where the brain receives only sparse, noisy data – what cognitive science calls the "poverty of the stimulus" – it must generate a rich, coherent simulation of the world, not unlike a quantum system resolving a superposition into a singular outcome. The mind doesn't merely interpret reality; it constructs it, selecting among probabilistic pathways much like a quantum measurement does. Thus, the mental world – our thoughts, dreams, and imagined futures – may be understood as a product of the same efficient, probabilistic rendering that governs particles and waves. In this view, consciousness becomes not just an emergent phenomenon, but a quantum renderer in its own right, turning incomplete data into the full experience of being.

Rendering Consciousness

The physical basis of consciousness

Schrödinger's intellectual curiosity extended far beyond the equations and enigmas of quantum physics; the profound mysteries of consciousness captivated him as well. In the tumultuous 1940s, as Europe was embroiled in war, Schrödinger sought refuge in the serene landscapes of Ireland, where he joined the Dublin Institute for Advanced Studies. Amid the rolling green hills and the whispers of ancient Celtic myths, Schrödinger turned his formidable mind to the enigma of what it means to be alive and aware.

His musings about consciousness ventured into uncharted territory and challenged the conventional boundaries between the mind and the physical world. He envisioned a reality in which consciousness and the material universe were not separate realms, but intimately interconnected, and potentially related through the very principles of quantum mechanics. Schrödinger speculated that consciousness might not emerge simply from complex neural interactions, as traditionally believed, but could have roots in the quantum realm itself. He pondered whether the unified nature of conscious experience – the seamless flow of thoughts, feelings, and perceptions that define our subjective reality – might echo the strange coherence of quantum states, where particles can exist in multiple forms at once, bound together across space and time.

In Schrödinger's view, the brain was a sophisticated biological machine, but also a potential quantum system, a complex dance of entangled particles that might underpin the continuity of conscious experience. This idea was revolutionary in an era dominated by classical physics, which viewed the world in terms of solid, separate objects that followed predictable paths. Schrödinger's suggestion that quantum mechanics could play a role in consciousness was met with skepticism, as many of his contemporaries found it difficult to reconcile quantum physics' unpredictable, probabilistic nature with the human thought's ordered, purposeful nature. Nonetheless, fascinated by the potential parallels, Schrödinger was undeterred even as he recognized the vast chasm between theory and observable reality.

Schrödinger's exploration of consciousness was not limited to the realm of physics; it also ventured into the philosophical, where he drew inspiration from Eastern philosophies like Vedanta, which proposes that all individual consciousnesses are facets of a single, universal mind. Schrödinger saw a harmonious resonance between these ancient teachings and his scientific insights, which suggested that individual consciousness might not be an isolated phenomenon, but part of a larger, interconnected whole. It was as if he glimpsed through the veil of ordinary perception and sensed a deeper truth that united the observer with the observed, the mind with the cosmos itself.

Ultimately, Schrödinger did not craft a definitive physical theory of consciousness, but his speculative ideas laid intriguing groundwork for future explorations. Today, as scientists probe theories like the quantum brain hypothesis, which explores potential relations between quantum mechanics and cognitive processes, Schrödinger's early thoughts stand as a testament to the unresolved mysteries at the nexus of physics and the mind. His legacy is not confined to the equations and paradoxes that bear his name; it also includes his bold willingness to transcend the boundaries of established science, gaze beyond the tangible, and confront the profound, unanswered questions of existence that continue to elude our grasp.

Integrated information theory

Integrated Information Theory (IIT) is one of the most daring and intriguing theories devised to unravel the puzzle of consciousness. Developed by neuroscientist Giulio Tononi at the University of Wisconsin-Madison, IIT posits that consciousness emerges from a system's capacity to integrate information in a unified and irreducible way. Consciousness is not a rare feature exclusive to humans or even living organisms, but a fundamental aspect that arises whenever information is intricately woven into a cohesive whole. This theory flies in the face of conventional wisdom, as it suggests that consciousness isn't merely a byproduct of neural complexity, but a fundamental property of systems that achieve a certain level of interconnectedness in processing information.

Five foundational axioms derived from the nature of conscious experience are at the heart of IIT. The first, Existence, asserts that consciousness is a fundamental reality; it undeniably exists because you can directly reflect on your vivid experiences. The second, Composition, states that consciousness is structured and constituted of interconnected elements. For example, when you look at a ripe apple, your perception isn't just a jumble of sensory data – it's a structured experience that integrates the apple's redness, roundness, and the gleam of light on its surface into a unified whole. Information is the third axiom, which emphasizes that every conscious experience is unique and distinct from all other potential experiences. The sight of a red apple is qualitatively different from hearing a birdsong or feeling the warmth of the sun on your skin.

The fourth axiom, Integration, is perhaps the most radical: it suggests that consciousness is a unified experience that cannot be parsed into independent parts without losing its essence. Your perception of the apple's color, texture, and shape aren't separate, but are fused into one seamless experience. Finally, Exclusion posits that consciousness is definite and specific and focuses on certain aspects of reality while it excludes others, much like a camera lens zooming in on a single object and blurring everything else.

Central to IIT is a concept known as Phi (Φ), a mathematical measure that quantifies the degree to which information is integrated within a system. The higher the Phi value, the more integrated and, consequently, the more conscious the system. Phi is like a scale that measures how much more the entire system knows compared to its parts in isolation. A high Phi value signifies a system in which information is interconnected so tightly that it cannot be simplified without losing essential characteristics. Not limited to brains, IIT suggests that any system – whether a human brain, a computer, or even an alien mind – could be theoretically conscious if it possesses sufficient integrated information.

Christof Koch, a prominent neuroscientist and Chief Scientist at the Allen Institute for Brain Science, has been instrumental in bringing IIT into the realm of empirical neuroscience. Koch, who has long pursued the neural correlates of consciousness, views IIT as a compelling framework to explain why certain brain regions, particularly those involved in thalamocortical interactions, are pivotal for conscious experience. Using transcranial magnetic stimulation (TMS) combined with EEG, researchers like Koch have observed that the brain's response to TMS is complex and widespread when a person is conscious and reflects high levels of integration. In contrast, during unconscious states like deep sleep or anesthesia, the brain's response is simpler and more localized, consistent with IIT's predictions of lower integration and, thus, lower consciousness.

Despite its appeal, IIT has not been without its critics. Anil Seth, a leading neuroscientist renowned for his work on conscious perception, questions whether IIT's mathematical approach can truly capture the subjective nature of consciousness – what it feels like to see red, taste chocolate, or feel joy. Critics argue that while IIT quantifies the integration of information, it may fall short in its ability to address our inner lives' subjective richness and complexity, known as qualia.

Still, IIT continues to push the boundaries of our understanding and proposes a universe in which consciousness is not confined to biological entities, but could be a widespread phenomenon wherever integrated information reaches a critical threshold. This revolutionary perspective opens up the tantalizing possibility that consciousness could take on diverse forms across different types of systems, from advanced AIs to complex networks that we have yet to discover.

Global workspace theory

In his quest to unravel the enigma of consciousness, cognitive neuroscientist Stanislas Dehaene guides us through the intricate labyrinth of the mind with the precision of a scientist and the inquisitiveness of an explorer charting unknown territories. As the head of the Cognitive Neuroimaging Unit in France, Dehaene has dedicated decades to peering into the brain's hidden workings by employing such advanced tools as brain imaging and computational modeling to reveal how we become aware of our thoughts and surroundings.

Dehaene's work has revealed that consciousness is not a monolithic process, but rather a dynamic symphony of neural activity in which different regions of the brain harmonize in a complex dance to create the vivid tapestry of conscious experience.

One of Dehaene's key contributions is the "global workspace theory", a concept that suggests that consciousness arises when information from various brain regions is broadcast across a vast neural network, similar to a spotlight illuminating the stage in a dark theater. In this model, the conscious mind operates like a broadcasting system, selecting and amplifying relevant information and sharing it with other parts of the brain for decision-making and action. It's as though the brain can use this intricate network to focus a beam of awareness on a specific thought or perception and bring it to the forefront while other processes continue behind the scenes, unnoticed.

Dehaene vividly illustrates this with experiments in which participants are shown images at the threshold of visibility – barely perceptible flashes that trigger neural activity, but often fall short of conscious recognition. By adjusting these images' timing and clarity, researchers can pinpoint the moment that a stimulus breaks through to consciousness, as it lights up brain scans like fireworks on New Year's Eve. In one notable experiment, Dehaene and his team employed functional MRI and magnetoencephalography (MEG) to track brain activity as participants processed words and images. When a word flashed too quickly to be perceived consciously, only localized brain regions lit up, like isolated sparks in a night sky. However, when the word was presented long enough to be recognized consciously, a cascade of neural activation swept across the brain and connected the visual cortex to areas involved in language, memory, and decision-making. Dehaene calls this phenomenon "neuronal ignition", which demonstrates that consciousness involves not only passive perception, but a comprehensive integration of information that enables complex thought, planning, and reflection.

Dehaene's research also delves into "masked priming" experiments, in which stimuli are hidden behind distracting images. For instance, a number might be flashed so fleetingly that the conscious mind misses it, followed by a masking image that obscures the initial stimulus. Yet, even without conscious awareness, the brain reacts. When participants are asked later to make quick numerical comparisons, the masked number influences their responses, which reveals that their subconscious minds processed the hidden information. This phenomenon underscores the dual-layered nature of cognition, where the subconscious mind processes a vast array of data while only a select fraction emerges into conscious awareness to guide decisions and actions.

Dehaene believes that consciousness can be quantified and broken down into specific neural correlates that can be observed, mapped, and even manipulated. According to him, the brain is not an impenetrable black box, but a decipherable system in which consciousness emerges from the complex interplay of attention, working memory, and the global neuronal workspace. He envisions a future when a deeper understanding of consciousness could lead to breakthroughs in AI, the development of neuroprosthetics, and innovative treatments for disorders of awareness that extend the boundaries of what we know about the mind and its remarkable capacity to construct the vivid reality that we navigate each day.

Quantum consciousness

Roger Penrose, a celebrated physicist and mathematician known for his groundbreaking work in cosmology and general relativity, has also ventured into the mysterious realm of consciousness – a journey that sets him apart from mainstream neuroscientific thought. In collaboration with anesthesiologist Stuart Hameroff, Penrose developed the bold and controversial theory of "Orchestrated Objective Reduction" (Orch OR), which posits that consciousness arises not merely from neural activity, but from actual quantum processes within the brain.

Penrose's foray into the science of consciousness was motivated by his dissatisfaction with conventional computational models of the mind, which compare the brain to a sophisticated computer that processes information through networks of neurons. To Penrose, these models seemed inadequate to explain the profound mystery of conscious experience – particularly qualia, the deeply personal sensations like the rich redness of a sunset or the sharp sting of a headache. Penrose argued that such experiences are too complex and profound to be mere byproducts of mechanical processes; he believed that something fundamentally non-computational, intertwined with the universe's fabric, was at play.

Penrose's fascination with the intersection of physics and consciousness was sparked by his reflections on the nature of mathematics and human understanding. He observed that mathematical insights often arise as sudden, intuitive realizations – the 'Aha' experience, a flash of understanding that defies the linear, step-by-step reasoning of algorithms. This line of thought led him to Kurt Gödel's incompleteness theorems, which revealed that within any mathematical system, certain truths cannot be proven by algorithmic means, and thereby highlighted the limits of purely mechanical computation. Penrose speculated that human cognition, particularly in mathematics, might tap into a deeper, non-algorithmic aspect of reality.

Delving into the world of quantum mechanics, Penrose explored the murky boundaries between reality and observation, where particles can simultaneously exist in multiple states until observed, at which point they collapse into a single state. He proposed that consciousness could be related to this quantum phenomenon through a process he termed "objective reduction". In Penrose's view, when a quantum system reaches a critical threshold of instability, it collapses from superposition into a definite state, which potentially could spark conscious awareness as it integrates a spectrum of possibilities into a singular experience.

As Penrose recognized that quantum effects would need a suitable environment to play a significant role in the brain, he turned to Stuart Hameroff, who had been exploring the potential of microtubules – tiny cylindrical structures within neurons that support cell shape and facilitate intracellular transport. Hameroff suggested that microtubules might offer the quantum-friendly conditions necessary to sustain coherence in the brain's warm and dynamic environment. Unlike the brain's typical neural networks, these microtubules are both small and structured in a way that could support quantum coherence and potentially function as miniature quantum computers within neurons.

Together, Penrose and Hameroff developed Orch OR, which proposes that consciousness emerges from quantum computations within microtubules, orchestrated by the brain's broader neural activity. In their model, when coherence within these microtubules reaches a certain threshold, it triggers objective reduction, which results in a discrete conscious moment – a "spark" of awareness that transcends classical neural processing.

The Orch OR theory has elicited both admiration and criticism. Supporters laud its audacity in associating consciousness with the fundamental laws of physics and suggesting that the brain might harness the universe's most enigmatic phenomena to generate the rich spectrum of subjective experience. However, critics argue that the brain's warm, wet, and noisy environment is far from conducive to sustaining delicate quantum effects, which are typically observed under highly controlled conditions, such as near absolute zero in laboratory settings.

Undeterred by skepticism, Penrose often draws parallels between Orch OR and his earlier work on black holes and singularities – ideas that initially encountered considerable resistance, but eventually became pillars of modern physics. His willingness to explore the uncharted waters of consciousness and quantum mechanics reflects his broader conviction that understanding the mind fully might require us to rethink our understanding of reality itself.

Penrose's bold speculations extend the boundaries of the way that we conceptualize the mind and its relation to the cosmos. He invites us to entertain the possibility that consciousness is not merely an emergent property of neural complexity, but is fundamentally linked to the universe's deepest structures. Quantum effects may offer the key to understanding the connection between the immaterial mind and the physical body.

Whether or not Orch OR withstands empirical testing, Penrose's work continues to inspire a vibrant dialogue at the intersection of physics, philosophy, and neuroscience and urges us to explore the tantalizing idea that our awareness may be intricately woven into the quantum fabric of the universe.

Personal reflections

I believe that quantum mechanics is like a cosmic rendering engine that had a relatively simple task in the quiet days of the early universe, before the emergence of life. At that time, the universe was like a vast sea of uncollapsed wave functions – probabilities suspended in limbo, a sprawling ocean of potential outcomes that didn't need to resolve into concrete reality. In this primordial state, the cosmos operated much like a lazy video game renderer that only sharpens into focus when observed. It existed in a kind of low-power mode, content with managing the sparse, monumental interactions of star collisions and the graceful waltzes of black holes without the need to render every particle in detail.

However, all of that changed when life appeared on the scene. With the eventual advent of conscious beings came a new and unprecedented challenge for the universe's rendering engine: now, the cosmos had to account for intentional decisions, complex behaviors, and these beings' ability to reflect on both the past and the future. Suddenly, the universe was no longer simply rendering random physical interactions – it had to evolve to manage this newfound complexity. Actions were no longer random; they were deliberate choices informed by the layers of experience, knowledge, and evolution that consciousness brought into play.

The human mind in particular complicates this cosmic rendering process by crafting its own independent simulation of reality, one that mirrors and even extends the universe's own. There's an intricate dance that occurs in our brains – predicting outcomes, planning actions, interpreting others' thoughts and intentions. When a surgeon decides where to make a crucial incision or an engineer sketches the design for a spacecraft, those decisions ripple far beyond the mere interactions of particles. They influence reality itself, determined by centuries of human history, accumulated knowledge, and intent. For the universe's rendering engine, this means more than just tracking the physical interactions of matter – now, it must also account for the complex web of conscious decision-making and the entire causal chain of human evolution.

This added complexity presents a unique dilemma for the universe's algorithms: it must now render both the cosmos' inert matter and sentient beings' unpredictable, conscious choices in vivid, seamless detail. Take, for instance, the lunar rocks that lay untouched for billions of years, resting quietly on the Moon's surface. They remained in their probabilistic state until the moment when human astronauts reached out, scooped them up, and brought them back to Earth. That journey – those rocks now rendered in their fullness – wasn't just the product of random chance, but a result of human ingenuity, motivated by curiosity and an almost boundless capacity for technological progress. We, as conscious beings, are very rare in the vastness of the universe, and the cosmos, ever efficient, likely optimizes its rendering process by managing conscious decisions differently from the simple interactions of inanimate matter.

I speculate that consciousness – self-aware, reflective, and capable of intentional action – has woven itself into the very fabric of the cosmos. Our observation forces objects into existence; without it, they remain in a probabilistic haze. Our minds are like parallel universes, rich with counterfactuals, "what-ifs", and abstract ideas like the concept of "redness", things with no direct counterpart in physical reality. We don't simply observe the universe – we actively influence it – and thus create our own mental renderings that extend beyond the physical world, and constantly test the limits of what the universe will allow us to imagine and create.

Rendering Free Will

The "do operator"

The universe can be considered a vast, nebulous mist that only clears when a conscious observer decides to look or take action, suggesting that reality itself hinges on a conscious agent's presence to bring it into sharp focus. It's as if the cosmos waits on standby, with its full complexity and detail locked in potential, ready to be rendered the moment that someone takes that decisive step of observation or intervention.

This idea finds a striking parallel in the pioneering work of Judea Pearl, a leading figure in the field of causality who developed an ingenious tool to disentangle mere correlation from genuine causation: the "do operator". To appreciate the effect of Pearl's "do operator", imagine stepping into the bustling atmosphere of a hospital research lab, where scientists are wrestling with a familiar yet formidable challenge – determining a new medication's true effects on blood pressure. As the researchers meticulously sift through their data, they observe that patients who are taking the medication tend to have lower blood pressure. At first glance, this correlation might seem like a breakthrough, but as any experienced scientist knows, correlation alone doesn't prove causation. These patients could also be engaging in other health-promoting behaviors, like going to the gym more regularly or religiously following a heart-healthy diet, which thereby muddies the waters of their analysis.

This is where Pearl's "do operator" makes its entrance. It allows researchers to simulate an intervention – similar to a scientist stepping into the fray, donning a lab coat, and directly administering the medication themselves. By "doing" or actively implementing the intervention, they can observe the outcomes in a controlled and systematic manner, and ask the critical question: "If we intervene in this specific way, will we observe a direct drop in blood pressure?" This shift from passive observation to active manipulation is crucial; it transforms a tangled web of coincidental associations into a clear exploration of causality that provides a pathway to distinguish between factors that merely coexist and those that actually determine change.

In the grander cosmic context, conscious human decisions function like the "do operator" by turning passive observation into dynamic participation. Just as the "do operator" clarifies the true effect of interventions in scientific studies, conscious actions bring the universe into focus and collapse the hazy mist of potentialities into a defined reality. Whether it's a scientist adjusting variables in a lab or a person simply choosing to look at a distant star, conscious decisions carve out pockets of clarity in an otherwise ambiguous universe and actively affect the world that we observe and experience.

Do we have free will?

Arthur Schopenhauer, the brooding 19th-century German philosopher, was not one to merely ponder abstract ideas from the lofty heights of academia; he lived his philosophy with a fervor that spilled onto every page he wrote. Immersing himself in the murky depths of human motivation and free will, Schopenhauer crafted theories that continue to challenge our beliefs about autonomy and self-determination. Drawing inspiration from Eastern philosophies such as Buddhism and Hinduism, together with the works of Immanuel Kant, Schopenhauer wove a complex tapestry that suggested that we are far from the masters of our fates. His iconic assertion, "Man can do what he wills, but he cannot will what he wills," encapsulates his view that while we can act according to our desires, deeper, unconscious forces beyond our control dictate the formation of those desires.

The core concept of Schopenhauer's philosophy is the "Will" – a fundamental, blind, and irrational force that he believed propels everything in existence. From animals' simplest survival instincts to the intricate motivations of human actions, Schopenhauer saw the Will as the bedrock of all reality. He argued that our actions are not the result of rational deliberation, but are manifestations of this underlying force. To illustrate this, he pointed to everyday scenarios: imagine someone steadfastly resolving to eat healthily, but finding themselves reaching for junk food time and again despite their best intentions. They might have the freedom to choose what to eat, but they cannot will themselves to prefer a salad over a cheeseburger. For Schopenhauer, this craving isn't a failure of resolve, but rather a clear expression of the Will – a force that operates independently of conscious thought and challenged the conventional notion of free will.

Schopenhauer's ideas intriguingly resonate with modern cognitive psychology and neuroscience. In the 1980s, psychologist Benjamin Libet conducted experiments that seemed to echo Schopenhauer's insights and brought them into sharper focus. Libet asked participants to press a button whenever they felt like it while he monitored their brain activity. Remarkably, he discovered that the brain's readiness potential – a precursor signal indicating the initiation of voluntary movement – occurred milliseconds before the participants reported that they made the conscious decision to act. This suggested that subconscious processes trigger actions, while conscious awareness lags behind. If our choices are indeed rooted in subconscious processes, as Libet's work indicates, the traditional concept of free will as a product of conscious thought becomes tenuous and raises profound questions about moral responsibility and justice – concepts that often rest on the assumption that individuals are deliberate, free agents.

This realization complicates our views on accountability and justice. If our actions are determined largely by subconscious impulses influenced by biology and environment rather than conscious choice, then holding individuals accountable becomes ethically complex. Society's legal and moral frameworks are deeply rooted in the belief in personal responsibility, and free will underpins our concepts of punishment, rehabilitation, and justice. Without it, these structures may need to be fundamentally reimagined.

Many scientists remain skeptical of free will, and emphasize genetics, environment, and neurobiology's influence on human behavior. Neuroscientist Michael Gazzaniga, known for his work with split-brain patients – a procedure that severs the corpus callosum in epilepsy patients to prevent severe seizures from spreading between brain hemispheres – suggests that the conscious mind often simply rationalizes decisions made by the subconscious already. He argues that while our conscious mind constructs coherent narratives about our actions, behavior's true determinants lie deep within the brain's neural circuitry, often beyond the reach of conscious awareness.

Schopenhauer's insight, "You can do what you will, but not will what you will," captures a profound aspect of the human condition: Our actions are dictated by our desires, but those desires are born from forces outside our conscious grasp. This perspective challenges us to reconsider the nature of free will and invites reflection on its implications for personal responsibility, morality, and justice in a world where the conscious mind is not the sovereign architect of our actions, but rather a participant in the deeper, unseen currents of the will. Schopenhauer's philosophy – and the way he lived it – reminds us that while we strive for control and autonomy, the forces that shape our lives often exert their influence from the shadows of our subconscious.

In my view, the decisions and observations that render the universe – our choices – can be either conscious or subconscious; it really doesn't matter.

Compatibilism

The universe is like a vast cosmic clockwork – a grand assembly of gears and cogs, each a metaphor for every event, action, or thought, all turning in a precise, predetermined dance governed by the immutable laws of nature. This is the essence of determinism: The idea that every occurrence, from the sweeping motions of galaxies to the tiny choices that we make each day, is dictated by prior conditions and unbreakable rules. If we had perfect knowledge of all of these initial conditions and the laws that govern them, the future would unfold as predictably as the ticking of a clock. It's an orderly, neatly arranged view of the cosmos, but one that conflicts with our deeply ingrained belief in free will – the feeling that, as conscious beings, we have the power to make choices independent of past events or external forces. We like to think that, in any given moment, we could have acted differently if only we had chosen to.

However, compatibilism offers a middle path and suggests that determinism and free will are not necessarily at odds, but can coexist. It redefines free will not as the ability to have acted differently in an identical situation, but as the capacity to act according to one's desires, motivations, and intentions – even if these inner states are themselves affected by preceding causes.

This reinterpretation preserves the critical concept of moral responsibility, a cornerstone of our ethical and legal systems. Compatibilism argues that individuals can still be held accountable for their actions, as long as those actions stem from their own motivations, much like holding a driver responsible for the route that he takes as long as he was steering the wheel, even if a predetermined GPS laid out the road ahead. This reimagined notion of freedom retains the vital sense of personal agency and accountability, which is essential to uphold social norms, justice, and the moral fabric of society.

The roots of compatibilism can be traced back to influential philosophers like David Hume, who deftly argued that liberty and necessity (determinism) coexist because freedom means acting according to one's will without external interference. Hume likened this to a river flowing within its banks: while the river's path is determined, it still flows freely within those constraints.

Similarly, Thomas Hobbes considered that free will is the ability to pursue one's desires without physical restraint, which suggests that deterministic processes do not negate free action as long as individuals are motivated by their impulses. Hobbes, who witnessed the upheaval of the English Civil War – a time of civil unrest, political strife, and the clash between the monarchy and Parliament – crafted a rather grim view of human nature, in which he argued for a strong, centralized authority to prevent chaos. In his seminal work, *Leviathan*, Hobbes described society as a great artificial man, where sovereignty acts as the soul and rewards and punishments as the nerves that guide behavior. His portrayal of human life as "…solitary, poor, nasty, brutish, and short" was a direct reflection of the turbulent times in which he lived.

Contemporary thinkers like Daniel Dennett have explored further how understanding human cognition within a deterministic universe can enhance our appreciation of freedom. Dennett and others argue that as complex adaptive systems, humans can exhibit behaviors that feel genuinely free, even if they are ultimately determined by prior causes. Dennett frames humans as sophisticated organisms capable of making nuanced, context-sensitive decisions, which offers a functional autonomy that is consistent with deterministic principles.

Compatibilism also finds practical application in such fields as law, psychology, and neuroscience. For example, the legal system operates on the principle that individuals can be held responsible for their actions as long as those actions are voluntary and uncoerced, regardless of whether their motivations are influenced by prior causes. This is consistent with the way that courts assess responsibility: Not by speculating whether a defendant could have acted differently in some metaphysical sense, but by considering whether his actions were a product of his own desires and intentions. Similarly, in psychology and neuroscience, compatibilism acknowledges the significant roles that genetics, environment, and upbringing play in influencing human behavior, while still affirming personal responsibility.

Of course, compatibilism is not without its critics. Incompatibilists argue that genuine free will cannot exist in a deterministic universe and insist that true freedom requires the potential to act differently under the same conditions – something that determinism denies explicitly. Advocates of libertarian free will, who support a form of indeterminism, believe that certain actions transcend the deterministic web and allow for true choice and agency.

Despite these ongoing debates, compatibilism offers a potential framework for reconciling free will with a deterministic universe. By redefining freedom as the ability to act according to one's motivations and desires, it preserves the essence of personal agency and moral responsibility. It's like finding harmony in a cosmic symphony in which every note is meticulously scripted, yet the melody of our lives still resonates with the unmistakable sound of autonomy and choice. Embracing compatibilism entails recognizing the deterministic forces that affect our paths, while celebrating the agency that empowers us to navigate those paths in ways that feel distinctly our own.

Interestingly, determinism suggests that every stage that leads to our current state – from evolution to human culture – must have been observed and transformed from its probabilistic quantum potential into a definite outcome.

Where does our "will" come from?

Consider settling into a dimly-lit cinema, popcorn in hand, as the opening credits roll. The familiar dance of storytelling begins on-screen: a plucky underdog faces impossible odds, a sinister villain plots his downfall, and the quirky sidekick delivers a well-timed joke that elicits chuckles from the audience. This cast of characters is no coincidence – it's a reflection of the human psyche, as neatly laid out in Blake Snyder's storytelling manual, *Save the Cat*. Snyder distilled the art of narrative into a formula and mapped out the arcs that tug at our emotions and keep us on the edge of our seats. Every twist, every hero's rise, and every villain's demise mirrors the stories to which we're instinctively drawn. Further, while we're engrossed in the plot, we seldom pause to ask why these characters are the way they are. We don't question what motivates the villain or the sidekick's quirks; we accept them as part of the grand spectacle.

However, step outside the movie theater and turn that lens toward the real world, where the kaleidoscope of human personalities surrounds us. Despite our differences, we share an astonishing 99.9% of our genetic code. DNA alone doesn't account for what sets the leader apart from the follower or the bold from the reserved. Behavioral geneticist Paige Harden delves into the complexities of how our experiences affect us and notes how the fire of battle can forge resilience or the way that the trials of a difficult childbirth can elicit profound empathy. Yet even these powerful experiences only begin to scratch the surface of what makes us who we are.

Consider the case of a recent American president whose persona – a combination of unrelenting motivation, unshakeable confidence, and a knack for controversy – set him apart in ways that transcend mere genetics. Compare him to your neighbor, who finds contentment in a quiet life far from the glare of the spotlight. It's not simply upbringing or personal choices that influence these differences. Some people are naturally drawn to the limelight, compelled to lead and influence, while others find their joy in quieter roles, offer support, and thrive behind the scenes. However, how can genetic differences cause trait differences if we're all pretty much genetically the same?

The truth may lie in a concept as mysterious as it is profound: an epigenetic lottery, a cosmic roll of the dice, after conception, that assigns personality traits randomly and steers our destinies in unforeseen directions. It's as if the universe is scripting our stories, much like a screenwriter crafts a character's quirks and motivations. Whether someone is destined to command a room, find solace in solitude, or dispassionately hire and fire staff without lingering doubts or loss of sleep, may be an evolutionary strategy.

These differences in human traits have deep roots and stretch back to the time before the dawn of agriculture. As our ancestors transitioned from nomadic life to settled communities, they began to specialize, much like ants in a colony, in which each person played a distinct role, from religious leaders to laborers. This division of labor became the backbone of societal progress, but despite the varied roles that we've played throughout history, our genetic similarity remains strikingly high. The roles that bestow wealth and privilege often feel less like rewards for effort and more like the outcomes of an enigmatic lottery.

This randomness in the allocation of traits – as well as disparities in wealth, health, and available opportunities – raises profound questions about fairness, yet our politics often complicate the matter. As researchers Keltner and Haidt point out, "Conservatives [are] less supportive of the notion that luck is influential to success because the randomness it invokes challenges their belief that people's outcomes are deserved, while the notion of random chance contributing to success is consistent with the liberal worldview." If our talents, inclinations, and life events are assigned in part by chance, is it fair and just that some should claim more of life's rewards simply because of the hand that they were dealt?

Enter the philosophies of scholars like John Rawls and Thomas Piketty, who might argue that a just society recognizes these inequities and seeks to level the playing field. Rawls, with his "veil of ignorance", proposes that we design our social systems as if we could be born into any position within them, which prompts us to consider fairness from all perspectives. Piketty, who focuses on wealth inequality, would likely champion the idea that wealth – often the fruit of this lottery – should be redistributed to ensure fairness. Thus, progressive taxation emerges as more than a financial policy; it's a recognition that the spoils of life are not purely earned, but frequently bestowed by the whims of the cosmic dice roll, which necessitates a more equitable distribution of the opportunities and resources that affect our lives.

Sentiments

The word "sentiment" encapsulates the entire spectrum of human emotions, cravings, fears, motivations, and instincts, and weaves together the essence of what it means to be human. Traditional psychology often draws a firm line between sentiments and instincts by treating the latter as hardwired behaviors evolved for specific survival purposes – like the meticulous nest-building of birds or the precise migration of monarch butterflies. However, this perspective is somewhat restrictive. Instincts aren't simply rigid blueprints etched into our DNA; they have evolved to be flexible and responsive to changing conditions, both in animals and humans. All human sentiments can be seen as instincts determined by evolution to be adaptable and malleable through experience.

Consider the classic definition of instincts: typically, they involve a stimulus that elicits a predefined response – a behavior that unfolds automatically without conscious thought. Yet, this definition overlooks a critical element: the feedback mechanism that often fine-tunes behavior based upon outcomes. For example, take the behavior of spiders, often hailed as paragons of instinctive action. A spider's web-building is directed by its ganglia – clusters of nerve cells that respond to sensory inputs. However, even this seemingly rigid instinct exhibits surprising versatility. Spiders can adjust their web-building strategies in response to environmental factors such as space constraints, wind conditions, or the availability of prey. They may alter the size or stickiness of their webs to optimize prey capture, which demonstrates a fundamental form of learning and adaptation. Spiders also show habituation and gradually ignore repeated, harmless stimuli like a leaf brushing against their web, and even display rudimentary associative learning, in that they modify their hunting tactics based upon past successes or failures. This adaptability challenges the conventional wisdom that instincts are immutable.

If even spiders can exhibit behavioral plasticity, it's not a stretch to suggest that human instincts, too, are malleable, trainable, and able to be refined through feedback. Human sentiments – whether ambition, conformity, or the pursuit of social acceptance – are rooted in instinctual determinants. Our brains come equipped with circuitry designed to detect particular situations and trigger predictable responses, but these responses are intended to be fine-tuned over time through upbringing, cultural influences, and personal experiences. For instance, we might instinctively feel angry when wronged, but other instincts, such as social conformity or the environment in which we were raised, profoundly influence the way that we express or suppress that anger.

Preferences and biases are deeply embedded in our cognitive processes and subtly steer our decisions. Daniel Kahneman's *Thinking, Fast and Slow* explores the way that we are inherently inclined to favor certain cognitive shortcuts, such as loss aversion – the tendency to fear losses more than we value equivalent gains – or the sunk cost fallacy, where we irrationally cling to a failing course of action because of prior investments. These biases are not learned behaviors; they are ingrained elements of our mental architecture designed to help us make rapid decisions that would have been advantageous for our ancestors' survival.

Similarly, personality traits – openness, conscientiousness, extraversion, agreeableness, and neuroticism – are stable characteristics that influence the way that we engage with the world. (Goldberg) These traits do not shift on a whim; they are deeply ingrained aspects of our psychology. An introvert is unlikely to suddenly become the life of the party, just as a highly conscientious person will rarely abandon caution. These personality traits are not simply reflections of our environment, but are woven into our very being.

Emotions provide another vivid illustration of our inherent traits at work. From the joy that bursts forth on a sunny day to the instinctive recoil at the sight of spoiled food, our emotional responses are attributable to neural circuits honed over countless generations. Babies don't need to be taught to smile or laugh; they naturally respond to playful faces and surprises with joy and wonder, thus showcasing a set of hardwired responses in their developing brains. Even complex emotional states, like social anxiety or addiction, are manifestations of these innate traits influenced by genetics, epigenetics, and the unique experiences that affect our lives.

Motivations follow a similar pattern. While basic motivations like hunger, thirst, and survival are shared across species, humans possess additional layers of motivation, such as ambition, the desire for social status, and the need for others' approval. These motives propel us to act – whether to strive for a promotion or avoid social disapproval. They're not learned habits, but aspects of our nature that are deeply ingrained.

For example, studies have shown that even young infants display a preference for "punishers" – individuals who uphold social norms by reprimanding wrongdoers – which indicates an innate sense of justice that manifests early in life. (Bloom)

Ultimately, our sentiments – whether they are preferences, biases, personality traits, emotions, or motivations – are what make us uniquely human. They guide our behaviors, influence our social interactions, and define the trajectories of our lives. While these traits can be suppressed, redirected, or refined, they rarely disappear entirely. It is crucial to recognize that innate traits do not imply strict genetic determinism. Although our brain's architecture follows a genetic blueprint – semantic vectors embedded deeply in our DNA – a complex interplay of genetics, epigenetics, environment, and personal choice influences these traits' expression. As mentioned previously, as we all have largely the same DNA, variations in traits may result from an internal epigenetic "roulette wheel" that assigns different levels of personality traits even before birth. This underscores that human differences are not rigid genetic destinies, but are part of the rich, dynamic tapestry of human diversity.

Hacking the brain

The interplay between biology, behavior, and perception provides a fascinating lens through which to explore the profound connections between our genetic code and the instincts that govern life. Usually, a mouse darts away from the faintest whiff of a predator and avoids the openness of an unprotected field. These instinctual behaviors are intricately encoded within the mouse's DNA. Think of these traits as "behavioral blueprints", or more precisely, structured DNA embeddings

– intricate genetic instructions that affect the way that animals respond to their environment.

For instance, consider the extraordinary case of *Toxoplasma gondii*, a microscopic parasite with a goal as peculiar as it is macabre: to ensure its own reproduction within a cat's gut. The parasite achieves this by manipulating its host – often a mouse – in a chillingly effective manner. Under normal circumstances, a mouse avoids such predators as cats at all costs, motivated by its instinctual fear response. However, when infected with *T. gondii*, this fear vanishes as if a switch has been flipped. The once-wary mouse now behaves recklessly and wanders into harm's way. How does the parasite achieve such a feat?

One theory suggests that *T. gondii* exploits the mouse's DNA-encoded behavioral embeddings and targets the neural circuits responsible for instinctive fear. These embeddings, like semantic vectors in AI that encode meanings in language, serve as a biological "dictionary" to interpret environmental cues. The parasite doesn't create a new behavior in the mouse, but rather suppresses its innate fear response and effectively hacks into the DNA's pre-programmed framework. It's as if the parasite understands its host's genetic operating system and uses it to disable the survival mechanism that would normally keep the mouse safe.

This manipulation is astonishing in its precision. Rather than forcing the mouse to behave in ways that are entirely unnatural, *T. gondii* simply removes a critical barrier – the encoded fear of predators. The result is a mouse that continues to function normally in every other respect, but lacks the caution needed to prevent its ultimate demise. It's a masterstroke of evolutionary ingenuity that highlights how external agents can exploit the genetic pathways that govern instinctual behaviors.

The concept of DNA embeddings offers a profound framework for understanding such innate traits. These embeddings may serve as a universal genetic structure, a form of encoded language that affects the way that organisms respond to common threats like predators, scarcity, or unfamiliar environments. For mice, this might mean fleeing at the scent of a cat; for humans, it might manifest as the immediate burst of anger when faced with unfair treatment – a modern echo of primal survival instincts.

Even in humans, the notion of DNA embeddings could explain shared instinctive responses to universal experiences. Think of the uncontrollable surge of road rage when another driver cuts you off. This reaction might stem from an encoded vector that equates "injustice" with a trigger for aggression, an ancient survival mechanism repurposed for the complexities of modern life. Much like *T. gondii* manipulates mice, external forces – whether social, environmental, or parasitic – could hijack these genetic circuits and amplify or suppress our instinctual reactions.

T. gondii's behavior is a vivid reminder of the way that these encoded traits not only dictate behavior, but also reveal vulnerabilities in the biological framework. By disabling fear, the parasite underscores the existence of a DNA blueprint that governs the way that animals perceive and respond to their world. Understanding this genetic encoding opens a window into the biology of behavior and offers profound insights into the way that innate traits are not simply programmed, but can also be manipulated.

Why do we feel?

On a quiet hillside, the cool breeze rustles through the grass as you tilt your head back to gaze at a sky that stretches infinitely above, studded with stars. As the vastness of the cosmos washes over you, a profound sense of awe takes hold – something far more mysterious, more intricate than a mere series of chemical reactions firing in your brain. Philosopher David Chalmers has famously called it the "hard problem of consciousness." This is no ordinary scientific puzzle; it's a conundrum that has flummoxed philosophers and neuroscientists alike for centuries. It asks the seemingly simple, yet staggeringly complex question: How do the firing neurons and synaptic dances of our brains give rise to the rich, vivid tapestry of subjective experience? It's not only about identifying which neurons flicker when you feel joy or sadness, but about revealing why that flickering transforms into the warm glow of a sunset, the sharp zest of a lemon on your tongue, or the visceral thrill of contemplating the very stars that you're watching. This transformation, in which the cold, mechanistic processes of the brain conjure up a colorful, vibrant inner world, is much like trying to pin down a star in the sky – elusive, vast, and infinitely complex. It's akin to reaching for the cosmos with bare hands.

For instance, take the raw, everyday sensations that we often overlook: the comforting warmth of sunlight on your skin, the juicy burst of flavor when you bite into a ripe strawberry, or the lingering pang of a bittersweet memory. These experiences, known as qualia, form the core of our conscious lives and present one of the greatest enigmas of the mind. Yet, they are more than simply fleeting feelings – they are the very essence of what it means to be alive, which potentially offers clues to our connection with the quantum world. From an evolutionary standpoint, these feelings are not mere fleeting whims; they are the ancient compasses that guide us through the labyrinth of life, influence our survival instincts, bind us to others, warn us of dangers lurking in the shadows, and push us to explore, create, and endure. They are woven into the deepest recesses of our biology, and perhaps, into the very fabric of the universe itself.

What if your feelings are not merely reflections of your brain's workings, but a manifestation of the quantum world – a reflection of the complex, interconnected simulation unfolding within your mind. In the bizarre realm of quantum mechanics, particles exist in a state of superposition in which they are able to be in multiple states simultaneously, and through entanglement, they can influence one another across vast distances in an instant. Similarly, our emotions are not isolated or straightforward; they are layered, interwoven, often unpredictable, and profoundly interconnected. It's as if our minds are engaged in a delicate quantum dance, in which each feeling nudges the next, like ripples in a pond that spread and overlap and create patterns of unimaginable complexity.

In the practical world of science, researchers are increasingly intrigued by the possibility that quantum effects may be at play in biological systems, including the human brain. Take the avian compass, for instance – a bird's astonishing ability to navigate thousands of miles guided by Earth's magnetic field, which hints that quantum coherence influences how these creatures perceive their world. If birds can harness the quantum realm to find their way across continents, is it so far-fetched to wonder whether our perception of reality, colored by feelings and qualia, might also be underpinned by similar principles? These aren't just idle musings; they belong to the burgeoning field of quantum biology, where the lines between life and quantum mechanics are beginning to blur.

In fact, "feeling" could be the moment when the brain transitions into a state of quantum consciousness, in which all of our conflicting memories and sentiments are instantaneously evaluated and lead to decision-making.

Feelings are not merely byproducts of our neural circuitry, but fundamental elements of our mental rendering that straddle the boundary between the tangible and the ineffable. They influence our identities, constrain our choices, and propel us forward in the intricate dance of life. From the primal fear that kept our ancestors vigilant to avoid predators to the love that binds families and communities together, feelings have been our silent guides, steering the evolutionary course subtly written in our DNA, and perhaps, reflecting the hidden, entangled patterns of the universe itself.

Why do we make up stories?

The complex dance of human decision-making unfolds largely below the surface of our conscious awareness, deep within the labyrinthine circuits of our subconscious. It's a bit like the inner workings of an elaborate clock: gears turning and cogs meshing with precise coordination, yet all hidden from view. We aren't equipped to consciously monitor the activity of each neuron in our brains any more than we could keep track of every ant in a sprawling colony. Our conscious minds strive to be consistent with reality and create a unified and coherent narrative, but beneath this facade lies a tangle of subconscious decisions and impulses.

To maintain a semblance of rationality amidst this internal chaos, our brains have evolved a specialized mechanism known as the "interpreter". This system acts like a storyteller or a master of spin that crafts post hoc explanations and justifications for our actions and beliefs. You have an internal PR officer working tirelessly to present your every action in the best possible light by weaving together a coherent narrative from the often-conflicting impulses that influence you. Michael Gazzaniga describes this interpreter as a relentless hypothesis generator that churns out explanations for our perceptions, memories, and actions. It's like having a smooth-talking spokesperson in your head, spinning tales to make sense of the myriad, hidden motivations behind your behavior.

Consider the interpreter's handiwork when you suddenly crave a piece of chocolate. The impulse doesn't spring from a conscious decision – it's likely a neuron firing somewhere deep within your brain that triggers a desire for a quick energy boost. You grab the chocolate and take a bite. When someone asks why, your brain promptly delivers a story: "I needed a pick-me-up," or "It was just there, tempting me." These narratives feel perfectly reasonable, but they are often little more than plausible-sounding post hoc justifications for decisions that the subconscious machinery of your brain already made.

The interpreter doesn't simply rationalize actions – it actively affects and reinforces our beliefs, even when they rest on shaky ground. Our brains are masterful at spinning tales that convince us of our own autonomy and rationality, even though subconscious processes determine many of our actions. Gazzaniga points out that this process of confabulation – crafting and believing our own made-up stories – is ingrained so profoundly that no amount of logical analysis can fully shake the sensation that we are deliberate and purposeful beings.

Within this framework, language isn't simply a tool to convey objective truths about the world; it's also a social glue that connects us to others, enforces societal norms, and increases our chances of survival. We are notoriously unreliable when we explain our actions and motivations and often concoct reasons that sound good, but may not reflect the actual forces at play. Yet the true power of language lies in its ability to build relationships, foster community, and help us navigate the complex social landscapes of our lives.

This intricate blend of subconscious decision-making and narrative crafting is what makes us distinctly human. With our brains' uncanny ability to interpret and confabulate, they are constantly constructing a version of reality that helps us stay connected to the world and to one another, all while they mask the myriad unseen influences that guide our actions. Thus, while the interpreter in our heads may not always offer the whole truth, it serves an essential purpose, as it helps us make sense of our experiences, maintain a sense of agency, and carve out our place in the ever-shifting web of human existence.

Rendering our decisions

In quantum mechanics, observation is not a passive act; it's an active, transformative engagement with the fabric of reality itself. In the peculiar realm of quantum physics, light doesn't conform to a fixed form. It exists in a state of superposition, both wave and particle, a blurred duality hovering in uncertainty. The moment that you observe it, you're not just passively witnessing light; you're actively influencing it. Your very act of looking causes the wave function to collapse and pins down light's uncertain state into definite particles – photons. It's as if the universe, responsive to your conscious awareness and subconscious judgments, reshapes itself in real time and tailors a defined reality to your observation.

This intimate dance between the observer and the observed hints at a profound truth about our relationship with the universe. Our capacity to perceive and influence reality isn't a recent development; it's the culmination of billions of years of evolution encoded in our DNA. Passed down through untold generations, this genetic material acts like a living archive, a bridge to the past that connects us to the earliest life on Earth. Our DNA carries the distilled wisdom of countless survival strategies refined over eons to sharpen our senses, reactions, and cognitive abilities. Each decision that we make – whether determined by instinct, emotion, or curiosity – taps into this deep evolutionary legacy and guides the way that we perceive and interact with the world around us.

Beyond our immediate perception, the remainder of the universe exists in a state of potential, much like the hidden corners of a video game world that remain unrendered until the player approaches. Think of a game like Fortnite, where vast landscapes, hidden secrets, and unseen opponents exist in the game's code, but only materialize on the screen when a player's character draws near. Until then, these elements exist in a kind of digital limbo waiting to be brought into view. Similarly, in the quantum universe, the parts that lie beyond our immediate focus exist as a shimmering canvas of possibilities undefined and unobserved. It is only when we direct our attention to something that it assumes a concrete form.

This act of rendering isn't confined to scientists peering through microscopes or quantum physicists in laboratories; it's a universal human experience. Each time that we engage with the world – whether by gazing at the stars, addressing a complex problem, or simply marveling at a sunset – we're not passive spectators. We are active participants, co-creators of the universe who affect its very existence with each thought and observation. In this ongoing interplay, our minds do more than reflect reality; they help forge it and demonstrate a dynamic collaboration between the human spirit and the vast cosmos.

The mind is a prediction engine

As you drive along a winding mountain road, the thick fog clings to your windshield, obscuring everything beyond a few feet ahead. Your hands tighten around the steering wheel, and your eyes strain to pick out the faintest hint of the path before you. However, it's not just your eyes that are guiding you – your mind is working tirelessly, predicting the twists and turns before they come into view. This extraordinary ability to anticipate what lies ahead, drawn from past experiences and subtle cues, underscores the mind's role as a remarkable prediction engine. Far from being a passive receiver of information, the human mind actively constructs a representation of reality that operates independently from the outside world, yet remains profoundly in sync with it and enables us to navigate through the fog of uncertainty with uncanny precision.

The mind's predictive prowess extends well beyond the immediate and ventures into counterfactuals and imagined scenarios. It's as if our brains are constantly running a complex simulation, exploring endless "what ifs" that allow us to plan, strategize, and innovate. A chess grandmaster, hunched over the board in a dimly-lit room, darts his eyes and mentally sifts through countless possible moves. In those moments, he's not merely reacting to the game's present state; he's peering several moves into the future, weighing the outcomes of different strategies, in which each possibility branches out like a vast, unseen tree. This mental simulation is so vivid that it seems nearly tangible, as though the pieces are moving on their own, animated by the power of thought alone.

One of the critical insights into the mind's predictive nature comes from neuroscience experiments that explore how our brains process visual information. In one study, researchers used functional MRI scans to observe participants while they looked at incomplete images of familiar objects. Even with only partial information, the brain's visual cortex would light up, filling in the missing details based upon past knowledge – like recognizing a friend from a fleeting glimpse in a crowded street. This phenomenon, known as predictive coding, shows that the brain doesn't wait for the full picture to appear; it leaps ahead, taking advantage of patterns and expectations to construct a coherent understanding of the world, often before our eyes can confirm it. (Friston)

The mind's capacity to imagine scenarios that don't exist in reality illustrates its independent representation of the world further. Remember Schrödinger's cat, a thought experiment that plays with the idea of superposition in quantum mechanics. There's no real cat placed in a box with a vial of poison, yet the mind can grasp the concept effortlessly and visualize the cat as both alive and dead until observed. This ability to entertain the impossible isn't just a quirky feature of human thought – it's a powerful cognitive tool that allows us to rehearse potential actions, foresee consequences, and innovate solutions before we take any steps.

Research in cognitive psychology has shown that the brain's simulation capabilities extend into everyday decision-making. Experiments have demonstrated that when people are asked to imagine future scenarios, like planning a vacation or envisioning a conversation, the brain's default mode network – the same network involved in recalling memories – lights up. It's as though the mind flips through a mental Rolodex of past experiences, piecing together elements to create a plausible future narrative. This simulated reality isn't confined by the present moment; it's a rich, flexible space where possibilities are tested, revised, and refined.

Sports provide some of the most striking illustrations of the mind's predictive nature. Watch a tennis player preparing to return a serve, her eyes fixed on the ball as it rockets toward her. In those split seconds, her brain isn't merely processing the ball's speed and trajectory; it's predicting where it will land, how to position the body, and the precise angle needed for the perfect return. This rapid-fire mental simulation allows athletes to respond faster than conscious deliberation would permit – a skill honed through countless hours of practice and feedback. The brain learns to anticipate the future by constantly comparing its predictions with actual outcomes, fine-tuning its internal model with each swing of the racket.

The mind's independent representation of reality functions like a sophisticated virtual reality headset that overlays the physical world with a rich tapestry of predictions, memories, and possibilities. It doesn't just reflect what is; it projects what could be by creating a seamless blend of perception and imagination. This synchronization between the mental and physical realms is so precise that it often goes unnoticed, yet it is fundamental to the way that we interact with the world, as it guides our actions, informs our choices, and allows us to navigate both the present and the future.

As we continue to unravel the mysteries of the mind, it becomes increasingly clear that our brains are not simply reacting to reality; they are influencing it actively, moment by moment, through a continuous cycle of prediction, feedback, and adjustment. This ability to think beyond the immediate, to explore scenarios that don't yet exist, sets human cognition apart in an elegant dance between what is known and what might be, played out on the grand stage of consciousness.

The mind emerges from the activity of the brain

In the 17th century, amid the lively bustle of Europe's cities and the dim glow of candlelit studies, René Descartes proposed a revolutionary idea that would alter our understanding of human consciousness forever: the mind and the brain are fundamentally different entities. For Descartes, surrounded by stacks of manuscripts in his study, immersed in contemplation, the brain was a marvel of mechanical engineering similar to the intricate clockwork automata that captivated the people of his time – an elaborate network of nerves and fluids operating with the precision of a finely-tuned machine. Yet, Descartes saw the mind as something altogether different – an immaterial, non-physical realm of thoughts, emotions, and intentions that couldn't be reduced to the workings of matter.

Descartes famously articulated this notion through his concept of dualism, in which he argued that the mind, or soul, was a distinct substance that interacted with the body through the pineal gland – a small, mysterious organ nestled within the brain's depths. This view was radical, as it challenged the prevailing belief that the body and soul were inseparable. He envisioned the mind as a "ghost in the machine", an independent force capable of abstract thought, vivid imagination, and conscious choice – faculties that, in his view, couldn't be explained merely by the physical operations of the brain.

To test these ideas, Descartes conducted thought experiments that explored the nature of perception and consciousness. He questioned sensory information's reliability and pondered how the mind could experience the world if it were truly separate from the physical senses. This line of inquiry led to his iconic declaration, "Cogito, ergo sum" – "I think, therefore I am." For Descartes, the mind's capacity for doubt, reasoning, and self-reflection set it apart from the physical brain and affirmed the mind's unique, non-material existence.

Descartes' theory of dualism sparked a debate that has persisted for centuries and strongly influenced philosophical and scientific discussions on consciousness, free will, and human experience. While modern neuroscience has shed light on how the brain supports thought and emotion, Descartes' core question – what separates the mind from the brain – remains an enduring topic of inquiry.

Today, neuroscience paints a detailed picture of the brain's role in decision-making and likens it to a bustling control center where specialized regions collaborate to shape our choices. The prefrontal cortex (PFC), located just behind the forehead, functions as the CEO of this vast enterprise, in which its subregions manage specific tasks: the dorsolateral prefrontal cortex (DLPFC) strategizes like a chess master planning moves and managing memory, while the ventromedial prefrontal cortex (VMPFC) evaluates emotional inputs like a meticulous accountant weighing rewards against risks.

This intricate decision-making network extends beyond the prefrontal cortex. The orbitofrontal cortex (OFC) adjusts decisions based upon changing outcomes, much like a financial trader responds to market shifts. Meanwhile, the anterior cingulate cortex (ACC) serves as a quality controller, flagging errors and resolving conflicts to ensure that decisions are consistent with past experiences and lessons learned. Deeper within, the basal ganglia guide habits and actions by drawing on the accumulated wisdom of past successes and failures, much like an experienced coach steering us toward reliable strategies.

Emotions also play a critical role in this cognitive symphony. The amygdala, our brain's emotional radar, responds swiftly to perceived threats or rewards and, influences decisions with urgency. The insula integrates bodily states with emotional and cognitive processes to offer gut-level insights that can instantly influence our choices. Meanwhile, the hippocampus taps into a vast repository of memories and provides the contextual backdrop for our actions.

These brain regions choreograph a complex dance of information gathering, evaluation, emotional input, and action selection. Sensory data flow from primary sensory areas to higher-order regions like the parietal cortex and craft a detailed representation of the environment. The prefrontal cortex then weighs options, anticipates outcomes, and calculates risks. Emotional and motivational inputs from the amygdala, insula, and basal ganglia add personal relevance, while the hippocampus enriches the decision-making process with memories and context. Finally, the action chosen emerges as the product of countless neural interactions.

Despite the remarkable strides that we've made in neuroscience, I find it hard to believe that mental processes can be fully reduced to brain activity alone. The mind appears to stretch beyond the synchronized firing of the brain's 86 billion neurons. Each neuron, with its thousands of interconnections, forms a dense and intricate network and communicates in ways so rapid and complex that it feels like more is happening than current neurological models can explain. Roger Penrose has ventured into the realm of quantum mechanics to suggest that perhaps the brain functions on a quantum level – a scale far smaller and faster than classical neuroscience can grasp. Neurons are not just biological messengers, but participants in a quantum conversation that share information through channels yet undiscovered by modern science. This tantalizing possibility – that the brain is tapping into quantum processes – suggests that the mind is something greater than the sum of its physical parts: an independent simulation of reality. It hints at a mysterious bridge between the tangible brain and an immaterial realm of thought and consciousness waiting to be fully explored. Could this quantum connection unlock the deeper secrets of the mind, and reveal what consciousness truly is? We are only scratching the surface of this profound enigma, with far more to learn.

Rendering AI

AI Success Stories

AI's journey from rudimentary algorithms to groundbreaking, transformative technology reads like a tale of science fiction made real. One pivotal chapter in this narrative unfolded in 1997 when IBM's Deep Blue stunned the world by defeating Garry Kasparov, the reigning world chess champion. This wasn't simply a triumph of raw computing power over human intuition; it was a bold proclamation of AI's burgeoning potential to rival, and even surpass, human cognitive abilities. However, the true revolution came nearly two decades later, when DeepMind's AlphaGo stepped into the arena of Go – a game celebrated for its profound complexity and the creative, nearly intuitive insights it demands from its players. In 2016, when AlphaGo faced Lee Sedol, a legendary figure in the world of Go, the stakes were higher than ever. This was no mere battle of man versus machine; it was a moment that redefined the boundaries of what was thought possible. AlphaGo's moves were audacious, confounded experts, and revealed new depths to a game that humans had played for millennia. The victory wasn't just on the board – it was in the rewriting of Go's strategic landscape, which showcased AI's capacity for creativity that extended beyond human comprehension.

Yet, AI's aspirations have always aimed higher than just mastering games. One of its most significant real-world achievements came with DeepMind's AlphaFold, which addressed the longstanding scientific challenge of protein folding. Proteins, the molecular engines of biology, fold into intricate three-dimensional structures that are crucial to their function. Predicting these shapes from a simple sequence of amino acids is a daunting task similar to solving a jigsaw puzzle with billions of potential pieces and no guiding picture. In 2020, AlphaFold shattered expectations by predicting protein structures with astonishing accuracy and achieving in hours what could have taken years of laborious experiments. This breakthrough holds profound implications for fields like drug discovery, where understanding protein shapes is vital, and for unraveling diseases' molecular underpinnings, which promises to accelerate advances that could reshape medicine and biology.

AI's prowess has also expanded into the nuanced world of language, thanks to the development of transformer architectures. These neural networks excel at processing sequences of data, which makes them suited particularly well to handle text. Unlike early AI models that relied on rigid, human-crafted rules and limited datasets, transformers are designed to digest massive amounts of text from across the internet, where they soak up the patterns of language much like eager students immerse themselves in books, conversations, and articles. You learn a new language not by rote memorization, but by immersing yourself in the flow of conversations. This is how transformers operate, and it allows models like ChatGPT to generate text that is not only coherent and contextually relevant, but strikingly human-like in its flow and nuance.

These advancements' practical applications are vast and transformative. For example, ChatGPT has evolved into a tool that can address a wide range of tasks, from answering complex questions and drafting essays to engaging in seamless, natural-sounding dialogue. While its responses aren't perfect, as they're grounded more in statistical pattern recognition than true understanding, it represents a leap forward in the way that machines interact with humans. AI models like

ChatGPT are reshaping virtual assistants, redefining customer service, and enhancing educational tools by infusing them with a more human touch.

These leaps forward reveal AI's growing versatility and hint at a future when technology doesn't just perform tasks, but collaborates with us and enhances human capabilities in profound ways. Whether it's solving intricate scientific puzzles, mastering complex games, or engaging in conversations that feel genuinely responsive, AI stands as a powerful ally in our quest to explore, understand, and influence the world around us.

Neural networks

In the early days of AI, the perceptron – an early model of an artificial neuron that Frank Rosenblatt crafted in the late 1950s – captured researchers' imagination with its promise to mimic human learning. Celebrated initially for its ability to perform simple tasks like distinguishing light from dark images, the perceptron offered a tantalizing glimpse into a future when machines could replicate the workings of the human brain. However, this enthusiasm waned quickly when researchers encountered its limitations, particularly its inability to solve non-linear problems, such as the XOR problem. This sobering realization led AI researchers to turn away from neural networks and shift their focus to symbolic AI and logic-based systems. These so-called expert systems were designed to replicate human decision-making by codifying the complexities of human thought into intricate webs of rules and symbols.

One of the most ambitious projects of this symbolic AI era was Cyc, an endeavor that computer scientist Douglas Lenat spearheaded. Lenat envisioned creating a vast knowledge base that could reason like a human by compiling millions of facts and rules about the world. However, the task of teaching a machine every nuance of human common sense – from the simple fact that water is wet to the consistent expectation that the sun rises in the east – proved Herculean. Cyc was an audacious attempt to encapsulate human knowledge, but as Lenat and his team soon discovered, reality was far messier than any static rulebook could handle. Symbolic AI systems like Cyc proved to be brittle and easily overwhelmed by the unpredictability and subtlety of real-world scenarios. It was like navigating a bustling city with an outdated map – each uncharted detour left you hopelessly lost. Similarly, symbolic AI stumbled whenever it encountered situations outside its predefined database of rules, which highlighted the rigidity and fragility of this approach. (Mitchell)

This inflexibility underscored the need for a more adaptable method and paved the way for the resurgence of neural networks. Unlike the basic perceptrons of the past, modern neural networks could learn and adapt by training on vast datasets, which enabled them to recognize patterns and make decisions in ways that symbolic AI never could. Neural networks thrived where symbolic systems faltered, particularly in complex tasks like image recognition, where they processed information through multiple layers of interconnected nodes that mirrored the way that the human brain might recognize a familiar face in a crowded room. Rather than rigid rules, neural networks operated within a dynamic, fluid, mathematical landscape known as semantic space, where relations between concepts were mapped as vectors that captured the intricate nuances of human cognition.

From symbols to semantics

Consider a neural network that can understand words' deep semantic meaning – not just as a series of characters, but in relation to other words. That's word2vec. Developed in 2013 by a team of Google researchers that Tomas Mikolov led, word2vec represented a major leap in natural language processing. While symbolic AI worked well for some problems, when it faced the challenge of understanding the nuances of language, it fell flat. That's where word2vec shined – it didn't need human-crafted rules; it simply learned through vast amounts of text.

Specifically, word2vec operates by creating word embeddings, which are essentially dense semantic vectors that represent words based upon their context. If you were to train word2vec on a body of text, you'd find that it could capture fascinating relations between words. For instance, it would determine that "king" is to "queen" as "man" is to "woman", not because it was told this directly, but because it observed patterns in the way that these words are used. It learns that "Paris" relates to "France" in the same way that "Berlin" relates to "Germany." Moreover, it goes further – synonyms group together, and words with similar roles in a sentence, like verbs or adjectives, will cluster in the same region of vector space.

To me, the most captivating part is how word2vec elegantly demonstrates that neural networks can build a sort of semantic map – a vector space where the meanings of words are expressed through their relations with other words. This stands in stark contrast to symbolic AI, which relies on pre-programmed rules and can't adapt as easily to new data or contexts. With word2vec, we don't teach the machine to recognize meaning – it learns meaning on its own. If you visualize the vectors, you see fascinating structures: clusters of related concepts, distinct neighborhoods of words. Words like "apple" and "orange" fall near each other because they share context in that they're frequently discussed as fruits, not colors or companies.

One of my favorite examples that demonstrates this power involves simple arithmetic with words, something that symbolic AI couldn't dream of doing. By taking the vector for "king", subtracting the vector for "man", and adding the vector for "woman", word2vec reliably returns "queen" – a brilliant example of how these word vectors encapsulate deeper linguistic relations. For anyone who is dabbling in natural language processing, word2vec is a fun and mind-expanding playground, which shows how a few lines of Python can lead to models that truly understand language in a way that feels almost magical.

While word2vec has since been superseded by transformers, it's still a lot of fun to try it yourself:

https://code.google.com/archive/p/word2vec/

Can AI learn like a child?

You wish to teach a child to recognize the world around them. With traditional methods, you'd point to objects and painstakingly **label** them: "This is a cat", "That's a tree", "Here's a car". That's the foundation of supervised learning – feeding machines an enormous collection of labeled data. This was the approach that led to the revolutionary AlexNet in 2012, a neural network that competed in the ImageNet Large Scale Visual Recognition Challenge. ImageNet, a dataset of over 14 million images that span 20,000 categories, became the gold standard for training AI systems. By training on the labeled ImageNet images, AlexNet was able to correctly identify 85% of new images. However, this approach came at a cost – every image in ImageNet had to be manually labeled, an effort so labor-intensive that it often felt like chiseling a masterpiece out of stone one painstaking tap at a time.

Self-supervised learning (SSL) followed, a method that turned the process on its head. Rather than requiring exhaustive labeling, SSL allows machines to learn by discovering patterns within the data themselves, much like a curious child playing with a puzzle to determine how the pieces fit together. For instance, imagine showing a machine the sentence "The cat sat on the ____" and asking it to predict the missing word, or presenting it with a sequence like "The cat sat on the mat" and challenging it to guess what might come next. By taking advantage of these intrinsic patterns, SSL eliminates the need for labeled data, which makes training faster, less resource-intensive, and infinitely scalable. The same SSL technique can be used to pre-train image recognition systems by asking the machine to predict known, but deliberately masked, parts of an image.

Even advanced AI systems like ChatGPT owe their capabilities to SSL. During its training, ChatGPT processes vast amounts of text without any human annotations and learned the intricacies of grammar, syntax, and meaning purely from patterns in the data. Later, fine-tuning with human feedback polished its conversational skills, and reinforcement learning refined its ability to respond thoughtfully. This blend of SSL, supervised adjustments, and human collaboration has resulted in a model that feels natural and nearly human in its responses. While supervised learning, championed by datasets like ImageNet, laid the foundation for modern AI, SSL has ushered in a new era – one where machines learn like humans, guided by patterns, curiosity, and a motivation to make sense of the world.

Rise of the Transformers

The next revolution came in 2017 with the introduction of transformer architectures – a groundbreaking design that fundamentally changed the way that machines interpret and generate human language. Introduced by researchers at Google in their pivotal paper entitled "Attention is All You Need," transformers revolutionized AI by focusing on attention mechanisms. Unlike earlier models that processed data sequentially, transformers used attention to focus on multiple parts of the input simultaneously, similar to the way that a spotlight highlights the most relevant details of a scene. This allowed transformers to excel in understanding long-range dependencies in text and capture context with remarkable precision.

Three essential components: the encoder; the decoder, and the attention mechanism, are at the heart of transformer architecture. The encoder (essentially a neural network) processes input data into manageable segments and captures the subtleties within each. The decoder then generates the output, whether it's translating text, crafting dialogue responses, or performing another form of language processing. The real marvel lies in the attention mechanism, particularly "self-attention", which allows the model to assess the relation between every word in a sentence simultaneously and weighs each word's importance in relation to the others (also using a neural network). For example, in a complex sentence like "The cat, which was hiding under the bed, finally came out," the attention mechanism helps the model correctly link "came out" to "the cat" despite intervening words that might otherwise confuse less sophisticated systems.

This ability to consider all parts of the input simultaneously accelerates training and dramatically improves the model's contextual understanding. It's like conversing with a friend about a vacation that he took months ago, and then weaving that conversation back into the current dialogue. Transformers effortlessly maintain these threads and track context across extended exchanges – a stark contrast to earlier models that often produced disjointed responses when conversations stretched beyond a few turns.

Transformers have become the cornerstone of many AI applications and extend beyond language to fields like image recognition and protein folding. Vision Transformers (ViTs) apply the attention-based approach to visual data, which allows these models to grasp spatial relations within images. Similarly, DeepMind's AlphaFold takes advantage of transformers to predict proteins' three-dimensional structures from their amino acid sequences, which has revolutionized molecular biology and offers immense potential for drug discovery and medical research.

These models have also permeated everyday technology and enhanced tools like Google Translate with the ability to manage idiomatic expressions, cultural nuances, and complex context shifts. In text generation, transformers power chatbots and virtual assistants and enable them to engage in fluid, natural conversations that can even capture elements of humor or empathy. Models like ChatGPT, built on a transformer architecture, vividly illustrate this progress, as they maintain a dialogue flow that feels uncannily human and adeptly navigates the twists and turns of conversation with a fluency once beyond AI's reach.

This journey from the rigid, rule-bound frameworks of symbolic AI to the adaptive, nuanced reasoning of transformers epitomizes AI's broader evolution. We've moved from the early days of navigating with static, inflexible maps to sophisticated systems similar to a GPS that not only guides us along set paths, but dynamically adapts as new information emerges. The shift from perceptrons' modest capabilities to transformers' expansive potential represents a monumental leap and is altering our understanding of machine intelligence and expanding the horizons of what AI can achieve.

Intelligent Agents

Can AI transcend its role as a mere tool and becomes a trusted collaborator able to address complex problems with minimal human oversight? This vision is rapidly becoming a reality because of the rise of AI agents – autonomous systems that can independently understand goals, devise plans, and execute tasks. Major players in AI, including OpenAI, Google DeepMind, Anthropic, Microsoft, and Meta, are leading this revolution and designing agents poised to reshape industries and redefine the way that humans interact with machines.

Take OpenAI, for example. Through innovations like AutoGPT, they are building systems that don't simply answer questions, but proactively take initiative. Consider an AI agent tasked with managing a marketing campaign. It wouldn't just draft content – it would conduct market research, analyze performance metrics, and refine its strategy, all without human micromanagement. Google DeepMind is advancing similar efforts with projects like AlphaCode, an agent able to write and refine software code autonomously, and MuZero, which masters games and strategies without being explicitly taught the rules. These breakthroughs highlight how agents can manage tasks that were once the sole domain of human experts.

Meta is extending the boundaries with projects like CICERO, an AI able to negotiate, strategize, and collaborate with humans in complex games. Meanwhile, Anthropic, founded by former OpenAI researchers, focuses on creating "aligned" agents – systems that prioritize safety, transparency, and ethical decision-making, while ensuring that they act responsibly in sensitive scenarios. Taking advantage of its partnership with OpenAI, Microsoft is embedding agent technologies into its products, which allows businesses to seamlessly automate workflows, analyze data, and enhance customer experiences.

Moreover, AI agents are not confined to the digital realm. Companies like Boston Dynamics and Tesla are integrating agent technology into robotics, which enables machines to adapt to physical environments in real time. Tesla's Full Self-Driving (FSD) program exemplifies this, in which autonomous agents analyze millions of data points to navigate roads safely and efficiently. Agents also show their collaborative potential in disaster response. Consider a coordinated team of agents: one analyzes satellite images to identify affected areas; another directs drones to deliver medical supplies, and a third optimizes supply chains – all in real time. DARPA is actively exploring this kind of multi-agent system for military and humanitarian applications.

The transformation extends to everyday life as well. In education, agents like Khanmigo, developed by Khan Academy, act as personalized tutors that adapt to each student's learning pace. In healthcare, systems like IBM Watson Health help doctors by analyzing patient data and recommending treatments. Retailers such as Amazon are deploying agents to optimize supply chains and customer interactions, while platforms like Zapier automate workflows and turn once-tedious tasks into seamless processes.

However, with great power comes significant challenges. As agents assume high-stakes roles, it becomes critical to ensure their ethical behavior and transparency. Companies like OpenAI and Anthropic are actively addressing these concerns by embedding alignment and safety mechanisms into their designs. Governments and institutions are also stepping in and creating regulatory frameworks to guide this rapidly advancing technology.

Ex Machina

In the 2015 film *Ex Machina*, we are introduced to Ava, an eerily lifelike robot housed in a secluded, high-tech research facility that feels more like a luxurious prison than a laboratory. The setting is remote, hidden in the wilderness and surrounded by mountains and dense forests, which underscores the isolation of Nathan, the reclusive tech mogul who created Ava, and Caleb, the unwitting programmer brought in to test her capabilities. As Caleb interacts with Ava through a series of Turing test-like sessions designed to assess whether a machine can exhibit human-like intelligence, Ava shows signs of awareness and a keen, calculating understanding of the human psyche. She flirts with Caleb, deceives him, and plays on his emotions, which gradually reveals that she is not just an advanced machine, but a master manipulator who exploits her understanding of human desires and fears to outsmart her creators and ultimately escape her confines.

This raises the tantalizing question: Does Ava's ability to manipulate and outmaneuver humans make her truly "artificially intelligent"? The answer is a resounding yes. After all, intelligence is not merely the capacity for computation or data processing, but involves a deeper, more nuanced grasp of the social fabric that binds us. Ava's actions suggest a sophisticated intelligence beyond rote responses or programmed algorithms. She mirrors others' internal emotional states, adapts her behavior based upon the reactions that she observes, and devises a strategy to free herself – a goal that she pursues with the kind of cunning and foresight that we typically attribute to humans alone.

Consider it this way: researchers often test animals for signs of intelligence in behavioral experiments by observing their ability to solve puzzles or navigate complex environments. Take the famous 1950s experiment with crows, in which scientists placed food at the bottom of a deep tube, and the crows had to determine how to retrieve it. They used tools – twigs and leaves – and bent them into hooks to pull out the food. This wasn't just a learned behavior, but an insightful use of their environment to achieve a goal. Ava operates on a similar level of intelligence, but with an added layer: she understands her environment and its humans' minds. She manipulates, predicts, and ultimately outwits them, thereby demonstrating a profound grasp of emotional and social dynamics.

Ava's ability to deceive Caleb and Nathan isn't just a clever plot twist; it's a display of AI that challenges our assumptions about what it means to be "intelligent". Suppose that a machine can mimic human behavior and understand and manipulate human emotions to serve its ends. In that case, it's not just following a script – it's thinking, strategizing, and, in a sense, living. Ava's intelligence isn't merely a simulation; it's a profound reminder that AI is as much about understanding the heart as it is about the mind. This is why Ava's story is so compelling: She doesn't just pass the Turing test; she transcends it and crosses a line that leaves us wondering where the boundary between human and machine intelligence truly lies.

Guardrails

AI has advanced a long way and evolved from basic algorithms into powerful tools capable of learning, adapting, and addressing challenges once thought to be the exclusive domain of human ingenuity. However, as we stand on the precipice of a new era in AI, it's becoming increasingly clear that the journey isn't simply about crafting smarter machines. The true challenge ahead is to create AI that is consistent with our values, serves the common good, and earns everyday people's trust. We need AI that doesn't just dazzle with its capabilities, but profoundly resonates with humanity by embodying ethics, empathy, and a genuine understanding of the human condition.

Think of the training process for most AI systems today: they're immersed in vast oceans of text gleaned from the far reaches of the internet, books, articles, and other written sources. It's as if these AI models are set loose in a colossal, dimly-lit library, racing through its towering shelves at lightning speed, absorbing and processing endless streams of words. While undeniably impressive, this approach often feels like a shortcut – a convenient way to sidestep the deeper, messier complexities of human nature. AI reads the script of our lives without ever seeing the full play performed, and thus misses the emotional subtext and the raw human experiences that give our words their true meaning.

Consider an AI model trained on millions of news articles. It can swiftly identify patterns, detect recurring conflicts, and spot economic and political themes, yet it lacks the capacity to feel the urgency of a breaking news story, the quiet resilience of a community rebuilding after a disaster, or the anger simmering behind a scathing editorial. For AI, words like "love", "fear", and "betrayal" are just data points, handled with the same clinical dispassion as "statistics" or "algorithms". The emotional weight, the lived experience behind these words, remains beyond its reach.

The heart of the matter lies in context – a vital element often missing in current AI systems. Humans bring a wealth of personal history, cultural understanding, and emotional nuance to every interaction. A simple phrase like "I'm fine" can carry countless meanings influenced by tone, setting, or the relationship between the speaker and listener. However, for AI, "I'm fine" is merely another input, processed without the subtext a human might instantly recognize as sarcasm, frustration, or a silent plea for help.

Therefore, the challenge is to move beyond the limitations of text-based training, to transcend the surface structure of language and engage with the full spectrum of human life – the vices, virtues, emotions, and unpredictable quirks that make us who we are. Creating AI that genuinely understands and reflects the richness of human experience requires blending technical prowess with a deep appreciation of the human spirit. It's not sufficient for AI to parse words; it must strive to comprehend the people behind them.

Looking to the future, AI's potential is as vast and varied as a night sky brimming with uncharted stars. It's about finding the right balance of guiding principles – similar to the way that a political party seeks to unify a diverse constituency. On the one hand, there's the vision of AI characterized by efficiency and meritocracy, which enhances productivity and extends the boundaries of what's possible. On the other, there's an increasing call for AI that prioritizes fairness, inclusivity, and the redistribution of opportunities to ensure that no one is left behind.

These contrasting visions mirror broader debates about AI's role in society. Should it be a relentless optimizer focused solely on output and performance, or should it be a compassionate ally dedicated to addressing systemic challenges and uplifting the marginalized? The answer likely lies in a delicate balance, a harmonious blend that marries efficiency with empathy.

An AI model that embodies this balance would champion personal agency while acknowledging the role of luck and circumstance, and step in to level the playing field when necessary. This AI would respect users' autonomy, uphold transparency, and safeguard personal data, thereby acting as a trusted partner in navigating the complexities of life.

This vision extends beyond mere technical ambition; it's a call to action for developers, policymakers, and society as a whole. To achieve it, we must collaborate to avoid the pitfalls of division and exploitation and ensure that AI serves as a force for good. We must guard vigilantly against AI's misuse for disinformation or to deepen societal divides. Ultimately, the goal is to harness AI's potential to craft a shared, hopeful vision of the future – one that inspires, unites, and protects against the dangers of unchecked ambition.

Artificial General Intelligence (AGI)

AI has made remarkable strides in recent years and captivated our imaginations with its rapid learning and ever-evolving capabilities. However, even with all of its advancements, today's AI systems remain far from replicating the nuanced and intricate intelligence of the human mind. At the core of this limitation is AI's inability to create an independent, simulated reality – a vivid mental model like the one that our minds construct naturally. We weave together sights, sounds, and emotions into cohesive experiences by drawing from a vast, interconnected tapestry of memories, common sense, and lived moments. This mental simulation reflects not only the world around us, but also anticipates what might happen next to guide our decisions and actions. In stark contrast, AI relies on static reservoirs of training data, which severely limits its grasp of deeper nuances such as cause and effect, counterfactual thinking, or complex "what-if" scenarios.

AI's limitations extend to fundamental aspects of human experience, such as the concept of time, space, and causality – domains where human common sense thrives. While we intuitively navigate the physical world with the ease of a finely-tuned GPS, creating mental maps, sequencing events, and grasping cause-and-effect relations, AI struggles with these basic concepts. It cannot remember specific events in meaningful ways, retrace steps along a familiar path, or recall the unique details of past encounters. This lack of episodic memory means that AI cannot form rich mental representations of experiences, learn from its mistakes, or adapt its behavior based upon past outcomes. It's as if AI views the world through a frosted lens, aware of the outlines, but blind to the finer details that define our everyday interactions.

A bird is perched on a branch, and within seconds you can intuitively recognize its form – head, wings, beak – and understand its likely behaviors. You know that it might take flight if startled or stay still if it feels safe. In contrast, AI frequently stumbles in such hierarchical categorization, and struggles to connect objects seamlessly with their attributes through observation and experience as we do. It cannot form mental maps or comprehend temporal dynamics, which makes even simple tasks like planning a journey or recounting a sequence of events challenging.

In the case of beliefs and evaluating information, AI falls behind even farther. Humans constantly assess the reliability of the information that we encounter by weighing sources and updating our understanding as new evidence arises. This adaptive skill allows us to effortlessly navigate a world of ever-changing information and filter fact from fiction. However, AI lacks this sophistication and relies often on surface-level patterns rather than deep comprehension. It can excel at specific tasks, but struggles with the complexities of real-world reasoning, such as predicting future outcomes, running hypothetical scenarios, or making strategic decisions that mirror human thought processes. The profound challenges that we face in creating artificial general intelligence (AGI) that can genuinely replicate human reasoning lie in these intricacies of the human mind.

Bold predictions from tech leaders like Elon Musk suggest that AGI that is able to surpass the most intelligent human could emerge within the next two years. Yet, as the tech world races toward this ambitious goal, a fundamental problem remains: the core concepts that would help us recognize AGI – mind, consciousness, emotions, free will, decision-making, planning, learning, and reasoning – still lack precise definitions. Without a clear understanding of these terms, proving the existence of AGI remains an elusive challenge.

Leading AI scientists such as Joshua Tenenbaum, Yoshua Bengio, Demis Hassabis, and Gary Marcus argue that to create machines with human-like intelligence, we must decode the principles that enable human cognition first. This includes our ability to learn rapidly from just a few examples, reason flexibly, and apply structured knowledge to novel situations. They envision AGI that mirrors the human mind's capacity for intuitive physics, social understanding, and causal reasoning – skills that come naturally to us, but remain elusive to current AI systems. They often draw on the analogy of a child learning about the world: children don't need thousands of examples to grasp basic concepts like gravity; they intuit them from just a few interactions. Similarly, they believe that AGI should generalize from limited data and learn robustly from minimal experience, much like a child experimenting with building blocks.

Other scientists, like Geoffrey Hinton and Yann LeCun, envision a future when AGI is achieved by expanding on the neural network architectures that have revolutionized AI already. Their work has been instrumental in developing the deep learning techniques that underpin modern AI systems like ChatGPT, but they acknowledge that these systems still fall short of true intelligence. They emphasize the need to integrate elements like attention mechanisms and symbolic reasoning to represent knowledge in a more human-like way. Moving beyond the current training paradigm of massive datasets is crucial, as it relies heavily on correlation rather than causation. They conceive of an AGI that can understand not just the "what" but the "why" and be able to generate causal explanations and predictions that transcend mere pattern recognition.

Today's large language models (LLMs), like ChatGPT, offer a glimpse into AI's potential, but also underscore current approaches' limitations. Trained on vast datasets of human language, these models draw from the words and ideas of millions of people, yet they don't truly understand the meanings behind those words. They can generate coherent text, but struggle with tasks beyond their training data, such as learning new concepts, reasoning abstractly, explaining their actions, or engaging in nuanced human behaviors that we take for granted –

such as gossiping, which involves tracking who said what to whom, understanding underlying sentiments and motivations, and assessing each piece of information's credibility. This complexity is part of what makes it so challenging to define AGI.

To reach the pinnacle of true intelligence, AGI must extend beyond mere data processing; it must develop an independent mind – a rich mental simulation with a nuanced representation of the world. For a self-driving car navigating a bustling city street, today's AI can manage the driving mechanics – staying in lanes, maintaining speed, avoiding obstacles – but the greater challenge lies in understanding the subtleties of human behavior and intent. When a pedestrian hesitates at the edge of a crosswalk, the car's AGI must predict, "Will she step into the road just as I approach?" Alternatively, consider a driver aggressively weaving through traffic: "Is he about to cut me off?" Even seemingly mundane actions, like a sudden swerve or an unsignaled turn, prompt questions about intent: "Why did she turn abruptly?"

Navigating three-dimensional space is straightforward for AI; it's a matter of calculating distances, velocities, and angles. However, understanding why a road worker has placed a traffic cone in a specific place requires insight into human reasoning, safety considerations, and the unspoken rules that guide daily actions. It's similar to reading a room or catching the nuance in a conversation – skills that we manage nearly instinctively, but that are profoundly challenging for AI. This is where AGI's real journey begins: moving beyond the rote execution of tasks to grasp human behavior's fluid and often unpredictable nature.

Although advanced, current AI systems still operate like tourists with a phrasebook in a foreign country – they may understand the words, but miss the cultural context that gives them meaning. To emulate human cognition's depth and richness, AGI must develop the ability to predict and interpret intentions, navigate social cues, and adapt instantly while maintaining a coherent and flexible model of its environment. It's not just about mastering the technicalities of a task, but engaging with the world as a genuinely sentient entity able to render the complex landscape of human thought and action with the same finesse that we do. This ambitious goal underscores the profound and exciting journey to create an AGI that genuinely thinks, learns, and interacts like a human.

Final thoughts

You step into the immersive world of a sprawling video game like Fortnite, where only the terrain directly around your character is fully rendered. Hills rise, buildings loom, and trees sway with lifelike detail as you explore. Yet, just beyond your immediate view, the game's universe remains in a state of suspended potential, coded, but not yet visible, waiting to be summoned into existence by your movement or gaze. This selective rendering isn't merely a clever trick to save processing power; it's a concept that strikingly mirrors quantum mechanics' fundamental nature. In the quantum world, particles exist in a state of superposition – a haze of possibilities – until they are observed and collapse into a definite state. It's as if the universe, like the game, is content to leave parts of itself unrendered, hanging in the balance until a conscious observer steps in and nudges potential into reality.

Our minds operate in much the same way, continuously simulating the world as we navigate our lives. This mental rendering is a dynamic, seamless process in which the brain acts like an incredibly sophisticated game engine. Neurons fire in complex patterns and weave together sensory inputs to create a coherent, vivid tapestry of experience. Just as a game engine stitches together graphics, physics, and players' actions to create an immersive world, the mind processes sights, sounds, and sensations, and blends them with memories and predictions to generate the fluid, ongoing simulation that we call consciousness. This capacity to render a personal version of reality is one of the mind's most remarkable feats as it turns raw data into a vivid mental landscape that guides our every step.

To render such intricate simulations, whether in the human mind or in advanced AI, requires immense computational power. This is where the frontier of quantum computing could play a transformative role. Just as GPUs revolutionized gaming by managing vast amounts of parallel data to create realistic visuals, quantum computers could offer the necessary horsepower to simulate the complex workings of the mind. Quantum bits, or qubits, can exist in multiple states simultaneously, which would allow quantum computers to process enormous amounts of information far beyond the capabilities of classical machines. This leap could bring us closer to the tantalizing possibility of creating an artificial mind that truly mirrors our own.

The quest to develop AGI is like an ambitious cosmic journey, one that attempts to unravel the intricate secrets of the way that our minds construct reality, anticipate the future, and navigate the complexities of human life. It's not merely about crunching numbers or sifting through mountains of data; it's about crafting systems able to render their own version of the world, much like the way that our minds create vivid, mental simulations. Your brain is like an ultra-sophisticated game engine that constantly weaves together sensory input, memories, and emotions to form a coherent narrative that helps you navigate daily life.

The real challenge for AGI is replicating this process – building an artificial system that can independently generate its own rich, detailed interpretation of reality. To do this, it would need the capacity to learn, adapt, and even simulate motivations and emotions to create a world filled with meaning and depth, much like the intricate tapestry woven by the human mind. It's a task that extends the boundaries of technology and understanding, as we try to mimic the extraordinary way that our minds render experience.

In this light, AGI wouldn't merely be a collection of algorithms; it would represent another mind, another simulation of reality rendered with the same depth and nuance as ours. The tantalizing possibility is that once we unlock the inner workings of our mental renderings – the way that we blend past experiences with present observations and future predictions – we might finally hold the key to crafting true AI. Such a creation would not simply process and respond to the world, but understand and interact with it in profoundly human ways, thereby bridging the gap between cold computation and the warm, complex fabric of conscious experience. With the right breakthroughs, we could edge closer to a future when artificial minds think, feel, and explore the universe together with us and share in the grand adventure of existence.

Timelines

Here's a detailed look at the significant milestones in the evolutionary journey from the dawn of life to the rise of human consciousness and the emergence of AI:

Origin of Life (Approximately 3.8 billion years ago)

- Life began in Earth's primordial oceans, likely from a soup of organic molecules that formed the first simple, self-replicating entities. These early life forms that may have resembled RNA-based replicators, marked the genesis of biology from chemistry.

Photosynthesis (Approximately 3.5 billion years ago)

- A transformative leap occurred when certain bacteria, like cyanobacteria, developed the ability to perform photosynthesis. Photosynthesis provides a new energy source by harnessing sunlight to convert carbon dioxide and water into glucose and oxygen. It oxygenated the atmosphere, which set the stage for more complex life forms.

Formation of Eukaryotic Cells (Approximately 2 billion years ago)

- Eukaryotic cells, characterized by a nucleus and organelles like mitochondria, evolved through a symbiotic relationship between primitive cells and engulfed bacteria. This event, known as endosymbiosis, allowed for greater cellular complexity and energy efficiency that paved the way for the evolution of multicellular organisms.

Multicellularity (Approximately 1 billion years ago)

- Cells began to band together and formed the first simple multicellular organisms. This cooperation allowed cells to specialize, and thus increased the efficiency and complexity of life. This milestone laid the groundwork for developing plants, animals, and fungi.

The Cambrian Explosion (Approximately 541 million years ago)

- The Cambrian Explosion was a rapid burst of evolutionary diversification, during which most major animal phyla appeared within a relatively short period. This explosion of life introduced new body plans, ecological niches, and complex behaviors, including the first predators and prey.

Colonization of Land (Approximately 500-400 million years ago)

- Life ventured out of the oceans and onto land. Plants were the pioneers, followed by invertebrates like insects and, eventually, vertebrates such as amphibians. This transition required significant adaptations, including developing limbs, lungs, and protective coverings.

Evolution of Reptiles and Mammals (Approximately 300-200 million years ago)

- Reptiles, including the ancestors of dinosaurs, diversified during the Permian and Triassic periods. The first mammals, small nocturnal creatures, evolved alongside dinosaurs and survived the mass extinction events that wiped out their larger relatives, which set the stage for the age of mammals.

Development of Primates (Approximately 65 million years ago)

- After the dinosaurs' extinction, mammals diversified rapidly. Primates, characterized by large brains, stereoscopic vision, and dexterous hands, evolved in tropical forests, where they adapted to life in the trees and developed complex social organizations.

Rise of Hominins (Approximately 7-6 million years ago)

- The human lineage diverged from the common ancestor with chimpanzees. Early hominins, such as *Sahelanthropus* and *Australopithecus*, exhibited bipedalism, which freed their hands for tool use and allowed for more efficient locomotion on the savannah.

Emergence of *Homo* Genus (Approximately 2.5 million years ago)

- *Homo habilis*, one of the earliest members of the *Homo* genus, began to use simple stone tools. The evolution of *Homo erectus* led to improved tool-making, control of fire, and migration out of Africa, which demonstrated an increased ability to adapt to diverse environments.

Development of Modern Humans (*Homo sapiens*) (Approximately 300,000 years ago)

- *Homo sapiens* evolved in Africa and exhibited advanced cognitive abilities, including sophisticated tool use, language, art, and complex social organizations. The development of abstract thinking, cultural transmission, and cooperation allowed humans to become the dominant species on Earth.

Agricultural Revolution (Approximately 12,000 years ago)

- The shift from hunter-gatherer societies to settled farming communities marked a pivotal moment in human history. Agriculture allowed populations to increase, cities to be established, and civilization to rise, which altered human societies and their environmental interactions fundamentally.

Industrial Revolution (18th-19th century)

- The Industrial Revolution brought about dramatic technological, economic, and social changes attributable to the invention of machines, the harnessing of fossil fuels, and the development of new manufacturing processes. This era accelerated humans' effects on the planet and reshaped daily life.

Digital Revolution and Rise of Artificial Intelligence (20th-21st century)

- The advent of computers, the internet, and digital technology has revolutionized communication, information processing, and access to knowledge. In recent decades, advances in AI and machine learning have begun to mirror and, in some cases, exceed human cognitive abilities in specific tasks, such as pattern recognition, game playing, and language processing.

Development of Artificial General Intelligence (Ongoing)

- Researchers are working to achieve AGI – systems that can perform any intellectual task that a human can. This quest involves creating machines able to learn, reason, and adapt across various tasks, which raises profound questions about the nature of consciousness, ethics, and the future relationship between humans and machines.

Further Reading

Adams, Douglas. *The Hitchhiker's Guide to the Galaxy*. Harmony Books, 1980.

Aspect, Alain, Philippe Grangier, and Gérard Roger. "Experimental Tests of Realistic Local Theories via Bell's Theorem." Physical Review Letters, vol. 47, no. 7, 1981, pp. 460-463.

Behe, Michael J. Darwin's Black Box: The Biochemical Challenge to Evolution. Free Press, 1996.

Bloom, Paul. Just Babies: The Origins of Good and Evil. Crown Publishers, 2013.

Bostrom, Nick. "Are You Living in a Computer Simulation?" Philosophical Quarterly, vol. 53, no. 211, 2003, pp. 243–255.

Bostrom, Nick. Superintelligence: Paths, Dangers, Strategies. Oxford University Press, 2014.

Buss, David M. Evolutionary Psychology: The New Science of the Mind. 5th ed., Routledge, 2015.

Calvin, William H. How Brains Think: Evolving Intelligence, Then and Now. Basic Books, 1996.

Carey, Susan. The Origin of Concepts. Oxford University Press, 2009.

Carroll, Sean. Something Deeply Hidden: Quantum Worlds and the Emergence of Spacetime. Dutton, 2019.

Chalmers, David J. The Conscious Mind: In Search of a Fundamental Theory. Oxford University Press, 1996.

Christian, Brian, and Tom Griffiths. Algorithms to Live By: The Computer Science of Human Decisions. Henry Holt and Company, 2016.

Clark, Andy. Mindware: An Introduction to the Philosophy of Cognitive Science. Oxford University Press, 2001.

Crick, Francis. The Astonishing Hypothesis: The Scientific Search for the Soul. Scribner, 1994.

Darwin, Charles. On the Origin of Species by Means of Natural Selection. John Murray, 1859.

Davies, Paul. The Demon in the Machine. Allen Lane, 2019.

Dawkins, Richard. The Blind Watchmaker. Penguin Books, 2006.

Dehaene, Stanislas. Consciousness and the Brain: Deciphering How the Brain Codes Our Thoughts. Viking, 2014.

Dehaene, Stanislas. How We Learn: Why Brains Learn Better Than Any Machine... for Now. Viking, 2020.

Dennett, Daniel C. Consciousness Explained. Penguin Books, 1991.

Descartes, René. Meditations on First Philosophy. Translated by Donald Cress, Hackett Publishing, 1993.

Doudna, Jennifer A., and Samuel H. Sternberg. A Crack in Creation: Gene Editing and the Unthinkable Power to Control Evolution. Houghton Mifflin Harcourt, 2017.

Edelman, Gerald M., and Giulio Tononi. A Universe of Consciousness: How Matter Becomes Imagination. Basic Books, 2000.

Edelstein, Linda N. Writer's Guide to Character Traits. 2nd ed.,
Writer's Digest Books, 2006.

Fodor, Jerry A. The Language of Thought. Thomas Y. Crowell
Company, 1975.

Ford, Martin. Architects of Intelligence: The Truth About AI from the
People Building It. Packt Publishing, 2018.

Friston, Karl. "The Free-Energy Principle: A Unified Brain Theory?"
Nature Reviews Neuroscience, vol. 11, no. 2, 2010, pp. 127-138.

Gamow, George. "Zur Quantentheorie des Atomkernes." Zeitschrift
für Physik, vol. 51, 1928, pp. 204-212.

Gazzaniga, Michael S. Who's in Charge? Free Will and the Science of
the Brain. HarperCollins, 2011.

Gazzaniga, Michael. The Consciousness Instinct: Unraveling the
Mystery of How the Brain Makes the Mind. Farrar, Straus, and Giroux,
2018.

Gödel, Kurt. "Über formal unentscheidbare Sätze der Principia
Mathematica und verwandter Systeme I." Monatshefte für Mathematik
und Physik, vol. 38, 1931, pp. 173-198.

Goertzel, Ben, and Cassio Pennachin, editors. Artificial General
Intelligence. Springer, 2007.

Goldberg, Lewis R. "The Structure of Phenotypic Personality Traits."
American Psychologist, vol. 48, no. 1, 1993, pp. 26-34.

Haidt, Jonathan. The Righteous Mind: Why Good People Are Divided
by Politics and Religion. Pantheon Books, 2012.

Harden, Kathryn Paige. The Genetic Lottery: Why DNA Matters for Social Equality. Princeton University Press, 2021.

Harris, Sam. Free Will. Free Press, 2012.

Hawkins, Jeff, and Sandra Blakeslee. On Intelligence. Times Books, 2004.

Hobbes, Thomas. Leviathan. Edited by J. C. A. Gaskin, Oxford University Press, 2008.

Hofstadter, Douglas R. Gödel, Escher, Bach: An Eternal Golden Braid. Basic Books, 1979.

Hume, David. An Enquiry Concerning Human Understanding. 1748.

Kahneman, Daniel. Thinking, Fast and Slow. Farrar, Straus, and Giroux, 2011.

Kant, Immanuel. Critique of Pure Reason. Translated by Marcus Weigelt, Penguin Books, 2003.

Keltner, Dacher, and Jonathan Haidt. "Approaching Awe, a Moral, Spiritual, and Aesthetic Emotion." Cognition and Emotion, vol. 17, no. 2, 2003, pp. 297–314.

Koch, Christof. The Feeling of Life Itself: Why Consciousness Is Widespread but Can't Be Computed. MIT Press, 2019.

Kurzweil, Ray. The Singularity Is Near: When Humans Transcend Biology. Viking, 2005.

Kurzweil, Ray. How to Create a Mind: The Secret of Human Thought Revealed. Viking, 2012.

Johnson, Phillip E. Darwin on Trial. InterVarsity Press, 1991.

Landauer, Rolf. "Irreversibility and Heat Generation in the Computing Process." IBM Journal of Research and Development, vol. 5, no. 3, 1961, pp. 183-191.

Lane, Hobson, Cole Howard, and Hannes Hapke. Natural Language Processing in Action. Manning Publications, 2019.

LeCun, Yann, Yoshua Bengio, and Geoffrey Hinton. "Deep Learning." Nature, vol. 521, no. 7553, 2015, pp. 436–444.

Lake, Brenden M., Tomer D. Ullman, Joshua B. Tenenbaum, and Samuel J. Gershman. "Building Machines That Learn and Think Like People." Behavioral and Brain Sciences, vol. 40, 2017, e253. Cambridge University Press, doi:10.1017/S0140525X16001837.

Lenat, Douglas B., and R. V. Guha. Building Large Knowledge-Based Systems: Representation and Inference in the Cyc Project. Addison-Wesley, 1989.

Marcus, Gary, and Ernest Davis. Rebooting AI: Building Artificial Intelligence We Can Trust. Pantheon Books, 2019.

Maxwell, James Clerk. Theory of Heat. Longmans, Green, and Co., 1871.

Michotte, Albert. The Perception of Causality. Translated by T. R. Miles and E. Miles, Methuen, 1963.

Miller, Stanley L., and Harold C. Urey. "A Production of Amino Acids Under Possible Primitive Earth Conditions." Science, vol. 130, no. 3370, 1959, pp. 245–251.

Minsky, Marvin, and Seymour Papert. Perceptrons: An Introduction to Computational Geometry. MIT Press, 1969.

Minsky, Marvin. The Society of Mind. Simon & Schuster, 1986.

Mitchell, Kevin. Innate: How the Wiring of Our Brains Shapes Who We Are. Princeton University Press, 2018.

Mitchell, Melanie. Artificial Intelligence: A Guide for Thinking Humans. Farrar, Straus and Giroux, 2019.

Newton, Isaac. Philosophiæ Naturalis Principia Mathematica. 1687.

Paley, William. Natural Theology: or, Evidences of the Existence and Attributes of the Deity, Collected from the Appearances of Nature. J. Faulder, 1802.

Pearl, Judea, and Dana Mackenzie. The Book of Why. Basic Books, 2018.

Penrose, Roger. Shadows of the Mind: A Search for the Missing Science of Consciousness. Oxford University Press, 1994.

Penrose, Roger. The Road to Reality: A Complete Guide to the Laws of the Universe. Vintage Books, 2005.

Phadke, Sharang. Unreal Engine 5 Game Development for Beginners: Kickstart Your Game Development Journey with Unreal Engine 5's Powerful Toolset. Packt Publishing, 2022.

Piketty, Thomas. Capital in the Twenty-First Century. Translated by Arthur Goldhammer, The Belknap Press of Harvard University Press, 2014.

Pinker, Steven. The Blank Slate: The Modern Denial of Human Nature. Viking, 2002.

Pinker, Steven. The Language Instinct: How the Mind Creates Language. Harper Perennial Modern Classics, 2007.

Rawls, John. A Theory of Justice. The Belknap Press of Harvard University Press, 1971.

Reich, David. Who We Are and How We Got Here: Ancient DNA and the New Science of the Human Past. Vintage Books, 2018.

Ridley, Matt. Genome: The Autobiography of a Species in 23 Chapters. HarperCollins, 1999.

Russell, Stuart J., and Peter Norvig. Artificial Intelligence: A Modern Approach. 4th ed., Prentice Hall, 2020.

Sadler, Matthew, and Natasha Regan. Game Changer: AlphaZero's Groundbreaking Chess Strategies and the Promise of AI. New in Chess, 2019.

Sapolsky, Robert M. Determined: A Science of Life Without Free Will. New York: Penguin Press, 2023.

Schrödinger, Erwin. What Is Life? Cambridge University Press, 1944.

Schopenhauer, Arthur. Essay on the Freedom of the Will. Translated by T. R. Miles and E. Miles, Methuen, 1963.

Seth, Anil. Being You: A New Science of Consciousness. Dutton, 2021.

Simler, Kevin, and Robin Hanson. The Elephant in the Brain: Hidden Motives in Everyday Life. Oxford University Press, 2018.

Snyder, Blake. Save the Cat!: The Last Book on Screenwriting You'll Ever Need. Michael Wiese Productions, 2005.

Sutton, Richard S., and Andrew G. Barto. Reinforcement Learning: An Introduction. 2nd ed., MIT Press, 2018.

Székely, Tamás, Allen J. Moore, and Jan Komdeur, editors. Social Behaviour: Genes, Ecology and Evolution. Cambridge University Press, 2010.

Szilard, Leo. "On the Decrease of Entropy in a Thermodynamic System by the Intervention of Intelligent Beings." Zeitschrift für Physik, vol. 53, 1929, pp. 840-856.

Vaswani, Ashish, et al. "Attention Is All You Need." Advances in Neural Information Processing Systems, vol. 30, 2017, pp. 5998–6008.

Venter, J. Craig. Life at the Speed of Light: From the Double Helix to the Dawn of Digital Life. Viking, 2013.

Virk, Rizwan. The Simulation Hypothesis: An MIT Computer Scientist Shows Why AI, Quantum Physics, and Eastern Mystics All Agree We Are in a Video Game. Bayview Books, 2019.

Watson, James D., and Francis H. C. Crick. "Molecular Structure of Nucleic Acids: A Structure for Deoxyribose Nucleic Acid." Nature, vol. 171, no. 4356, 1953, pp. 737-738.

Wheeler, John Archibald. "Information, Physics, Quantum: The Search for Links." Proceedings of the 3rd International Symposium on Foundations of Quantum Mechanics in the Light of New Technology, Physical Society of Japan, 1990, pp. 354-368.

Wigner, Eugene P. "Remarks on the Mind-Body Question." Symmetries and Reflections: Scientific Essays, Indiana University Press, 1967, pp. 171-184.

Wilson, Timothy D. Strangers to Ourselves: Discovering the Adaptive Unconscious. Belknap Press of Harvard University Press, 2002.

9 7 9 8 3 3 9 8 0 7 8 8 9